Edmund Beckett Grimpthorpe, Pliny Earle Chase

Astronomy Without Mathematics

From the Fourth London Edition

Edmund Beckett Grimpthorpe, Pliny Earle Chase

Astronomy Without Mathematics
From the Fourth London Edition

ISBN/EAN: 9783744727563

Printed in Europe, USA, Canada, Australia, Japan

Cover: Foto ©berggeist007 / pixelio.de

More available books at **www.hansebooks.com**

ASTRONOMY

WITHOUT MATHEMATICS.

BY

EDMUND BECKETT DENISON, LL.D., Q.C., F.R.A.S.

AUTHOR OF

THE RUDIMENTARY TREATISE ON CLOCKS AND WATCHES AND BELLS,
LECTURES ON CHURCH BUILDING, ETC.

From the Fourth London Edition.

EDITED WITH CORRECTIONS AND NOTES BY

PLINY E. CHASE, A.M.

———

NEW YORK:

G. P. PUTNAM & SON, 661 BROADWAY.

1869.

NOTE TO THE AMERICAN EDITION.

The great popularity of Denison's "Astronomy without Mathematics" has induced the American publishers to prepare an edition specially adapted for general circulation in the United States. The only changes which have been deemed advisable, are: 1. A few verbal alterations, in cases of accidental errors or inadvertences of style or statement. 2. The addition of an Appendix, to which references are made by foot-notes in the body of the work.

The care bestowed by the author in the preparation and arrangement of his materials, the general accuracy of his text, the simplicity of his explanations, his judicious presentation of interesting topics, and the valuable information embodied in the notes of the American editor, will commend the book to teachers as well as to private readers, and we believe that it will also be found worthy of a place in the libraries of professional astronomers.

PREFACE TO THE THIRD EDITION.

The sale of 3000 copies of this book in little more than a year, many of them to persons of more education than I originally contemplated, has induced me to enlarge it considerably, going rather deeper into the subject, and adding some explanations which I did not venture on before. The additions chiefly relate to meteors, nebulæ, and stars, the moon's acceleration and other disturbances, the tides, and the calculations for Easter in all ages; and a fuller account of the methods of weighing the sun, moon, and planets. The chapter on telescopes also has been greatly enlarged; for astronomy lives by them, and I do not know where a popular explanation of the theory of telescopes is to be found.

I keep to the plan of using as few 'diagrams,' and as few words, as will serve the purpose, because I am satisfied that such explanations are both easier to read and to remember—provided of course they *are* explanations (see p. 100 n). A book is not a lecture. There it may be prudent to say things several times over, in different ways, and to illustrate them as well as you can; for the hearers must not stop to think, or they will be left behind. I have not scrupled to draw figures wherever I thought they would be useful.

I repeat the warning of the former editions, that this book only aims at making astronomy as easy as it can be made if difficulties and the reasons of things are really to be explained, and not evaded in vague language which leaves people as ignorant as before. It is idle to suppose that anything can be learnt of astronomy as a science of causes and effects, without some study and power of thought, and some natural capacity for geometrical conceptions. But those who cannot always follow the reasoning may still read the results, treating the book as one of 'descriptive astronomy' only, though it is really an introduction to 'physical astronomy,' or the astronomy of causes and effects.

Though no mathematical knowledge is required of the reader, I do not profess anything so absurd as to rebuild the Newtonian system without mathematics. We soon come to a point in explanation where we must either stop and disclose no more, or else bridge over the chasm by adopting some simple result, or perhaps rather difficult calculation; such for instance as these two: that a sphere, but not a spheroid, attracts as if it were all condensed into its centre; and that the time of performing an elliptic orbit is the same as of the circle which contains it. All the arithmetical calculations here are founded on propositions as simple as these; and yet these require geometry, algebra, trigonometry, conic sections, and differential and integral calculus (or else other obsolete contrivances) to prove, though not to use them.

I have taken no small pains to avoid mistakes, but I cannot expect to have entirely succeeded; for even great astronomers occasionally commit them, from haste of writing or imperfect recollection; and sometimes they have to recant absolute mistakes of reasoning. I do not pretend to be an astronomer at all; but having been pressed into this service by a kind of accident, I have done the best I could for it.

E. B. D.

33 Queen Anne Street, W.,
February, 1867.

CONTENTS.

CHAPTER I.

The Earth.

CHAPTER II.

The Sun.

CHAPTER III.

THE MOON.

CHAPTER IV.

THE PLANETS.

CHAPTER V.

THE LAWS OF PLANETARY MOTION, COMETS, NEBULÆ, AND STARS.

CHAPTER VI.

On Telescopes.

ASTRONOMY

WITHOUT MATHEMATICS.

CHAPTER I.

THE EARTH.

For several thousand years people supposed that the world was a large platform, and that if you went far enough you would come to the edge everywhere, as you do at the sea shore. They thought the sun set in the sea and rose out of it; at any rate the old Greek poets said so, and invented wonderful contrivances for carrying the sun round or under the earth in the night from west to east again.*

But you may ask, does not David in the Psalms (18 and 39) speak of the 'round world?' No, he does not. There is no such word as 'round' in the original Hebrew; nor in the old Greek translation called the Septuagint, said to have been made by seventy-two learned men about three centuries before Christ; nor in the Bible version of the Psalms. Its appearance in the Prayer book is a curious result of successive translations. That version was translated from the Latin

---

* See Sir G. C. Lewis's 'Astronomy of the Ancients.'

1*

one called the Vulgate, which alone was used until the Reformation; and St. Jerome, who made the Vulgate Psalter from the Septuagint, translated the Greek word for 'the inhabited world' into that which had become the common name for the world with the Latin writers, who did know that the earth was round,* viz., *orbis terrarum.* And so Cranmer, 'whose image will be seen reflected on the calm surface of the liturgy while the Church of England remains,'† not unnaturally followed it, and translated almost literally *fundamenta orbis terrarum* into 'the foundations of the round world.' But the translators of the Hebrew Bible afterward did not.

It would have been contrary to the habit of the Bible to anticipate and reveal a scientific discovery, which men would make for themselves in time, and which was of no consequence to their religious faith and life. It is not contrary to the habit of the Bible, nor at all superfluous, to declare continually, wherever there is an opportunity (as we may say), that the sun and the moon and the stars, and the earth and all that is in it, did not grow of themselves, as some people fancy, but were created, or made out of nothing by the word of God. For all the science in the world could never prove that 'in the beginning God created the heaven and the earth:' we can only know that by revelation. 'Through faith (not science or observation and reasoning) we understand that the worlds were framed by the word of God, so that things which are

---

* Cicero, 'De Natura Deorum,' ii. 19, etc.
† Froude's 'History of England,' v. 391.

seen were not made of things which do appear' (Heb. xi. 3.) And all experience shows that men who disbelieve that, believe nothing else that is revealed. This has nothing to do with the proper interpretation of the particular words in which the successive acts of creation are described in the first chapter of the Bible. Nearly all learned men now agree at least in this, that the word translated 'day' there does not always mean a day of twenty-four hours, but may mean periods of enormous length, each ended by some marked division of time or epoch.

Thales, who was called one of the wise men of Greece, is commonly said to have discovered that the world is a round globe, about 600 B.c.* Consequently, men and water and all things stand all round it without falling off; and what we call 'upright' only means upright with reference to the surface of the earth or of water where we are, or in a line toward the centre of the earth; toward which all things fall or press; so that they fall in opposite directions here and in the Indian Ocean, which you may see is exactly opposite to the United States, in that model of the earth with the countries painted on it, which they call a *terrestrial globe.*

You may ask how we know all this. We know first that the earth is round, in some general sense of the

---

* Sir G. C. Lewis says there is no evidence in support of this tradition. It is of very little consequence; but on the other hand there seems a strong probability of the Chaldæans having known far more about astronomy than the mere roundness of the earth, ages before Thales. See a chapter on Chaldæan astronomy in Mr. Proctor's book on Saturn, and the late John Taylor's book on the Great Pyramid.

word, by finding that we can actually go quite round
it by sea or land in every direction, except where we
are stopped by ice or mountains. Secondly, we find
that the sea is nowhere flat, but rises everywhere like
a low round hill, so that the masts of distant ships are
always seen before their hull or body. So great plains,
which are *level* like the sea, are not flat like a table,
but rise visibly between two people at a distance, as
the sea rises between the ships. And such plains, and
level surfaces of water all over the world, rise the same
height in the same distance, viz., 8 inches in a mile,
and not 16 but 32 or 4×8 inches in 2 miles, 9×8
inches or 6 feet in 3 miles, and so on for a considerable
distance, exactly as it would upon a globe according
to geometrical rules. So that two very tall men standing 6 miles apart on a level plain can only just see
each other's heads with telescopes, and what we call
the 'visible horizon,' or boundary of sight by the level
ground or sea, is everywhere 3 miles from an eye 6
feet high. All this can only happen on a globe of a
certain size. And what we call *level* only means flat
when the surface is not large enough for the curve to
be distinguished. It means really the surface which
fluids take at rest, to which a plumb line is upright,
and which is equidistant everywhere from the centre
of the earth—subject to a small correction which you
will see at page 19.

Again we see the outline or shadow of the earth itself upon the moon in eclipses, and that is always
round, whatever part of the earth may face the moon
just then. Indeed, as eclipses were observed and pre-

dicted and recorded as important astronomical events long before distant voyages at sea were made, the roundness of the world was very likely first considered to be proved by them, though the notion of its being a platform may have been given up before. For a body which always casts a round shadow can be nothing but a globe, as you may easily see if you hold up things of other shapes before the sun in different positions.

Measuring the Earth.—After it was found that the earth is a globe, it was natural to try to measure it; but it was long before that could be done accurately. It may indeed be done approximately from the figures just now given; for it may be proved that if the earth is a globe, its diameter is to the distance of the visible horizon as that is to the height of your eye above the plain; which you will find gives 7920 miles, for a height of two yards and a distance of the horizon of 3 miles or 5280 yards. But this method admits of no great accuracy, and these figures are rather the result than the source of really accurate knowledge of the earth's size; for the rays of light near the ground are irregularly bent or refracted by the air, so that you do not in fact see straight, and cannot distinguish where the visible horizon for really straight lines of sight would be; and a very small error in the distance of the horizon will make a very large one in the size of the earth.

It has now been measured by other means which I will describe presently; and it is found to be 24,907 miles round the *equator ;* which is a circle round the middle of the earth at an equal distance from the

north and south poles. The poles are the two ends of
that imaginary axis round which the earth turns every
day. All circles round the earth and going through
the poles are called *meridians;* and so every place has
its own meridian, which runs from north to south, and
the sun crosses that circle at the noon or mid-day of
that place. All circles which divide any globe equally
are called *great circles*, because no greater can be
drawn. Any straight cut or section through a globe,
which does not divide it equally, makes a *small circle*.
The shortest road between any two places on a globe
is by the great circle which passes through them both:
hence comes what is called 'great circle sailing.' The
diameter of a globe is necessarily also the diameter (or
line through the centre) of every great circle; and you
should remember that the circumference of every circle
is very nearly 3⅐ of its diameter; that is, if the diameter is seven feet or miles the circumference may be
called 22; or more exactly, circumference = 3·1416
diameters very nearly; but no number of figures can
express the *exact* proportion. The *radius* is half the
diameter.

The greatest equatorial diameter is 7926·6 miles.
Some measurers of the earth make it nearly two miles
less than this through 104° east and 76° west longitude; but later calculations seem to make this doubtful (see p. 169). At any rate we may practically treat
the earth as round at the equator, and at all the small
circles parallel to it, which are called *parallels of latitude*.

The polar axis may be called 7899½ miles, or 500

millions of inches about a thousandth longer than our inch (a quite insensible difference), or 20 millions of the 'sacred' cubit of the Jews, according to Newton's estimate thereof (see p. 341). Sir J. Herschel remarks that the French metre, 39·371 inches, *the newest and worst measure in the world,* differs much more than that from the fraction it pretends to be, a 40 millionth of a far more uncertain quantity, the length of a meridian. A few minor scientific men fancy that by writing in French measures they can make the world follow them into adopting this inconvenient, inaccurate, *unstridable* measure, and its subdivisions. They only make the greatest part of the world not follow them, in another sense.*

The polar axis being thus about 26½ miles less than the mean equatorial one, the earth is not quite a globe or sphere, but what is called a *spheroid ;* which means something like a sphere. There are two kinds of spheroids, one flattened at the poles, and fatter round the equator, as the earth is, which is called an *oblate spheroid,* and is formed by turning an ellipse round its smallest diameter ; and the other, formed by turning an ellipse round its greatest diameter, is thinner at the equator, and drawn out at the poles, like an egg with two small ends, which is called a *prolate spheroid.* The spheroidicity of the earth or any other planet is usually called its ellipticity : which means the proportion between the difference of the two axes or semi-axes of an ellipse, and the greater of them ; or the proportion of EB to AC in the figure at p. 39.

---

* See Appendix, Note I.

Consequently every meridian of the earth is an ellipse, and not strictly a great circle ; though for ordinary purposes it may be called one, as the ellipticity, or the proportion of 26½ miles to 7926, is only one 298th, so little that you could not perceive it on any globe that could be made. An ellipse is· shorter than the circle containing it by very nearly half the ellipticity, so long as that is small : when it is not, the relation between them is complicated. Therefore a meridian is a 596th shorter than the equator. In giving the velocity of the earth and planets in their orbits, which are all elliptical, I shall treat them as circles for simplicity, as their ellipticity is very small, and nothing turns on the precise amount of the velocity.

As I have had to mention meridians and latitude, I had better explain at once what *latitude* and *longitude* are. The circumference of every circle may be divided into 360 equal parts, called degrees ; and again, every degree (1°) contains 60′ (minutes), and every minute (1′) contains 60″ (seconds), which have nothing to do with minutes and seconds of time. That is the way that parts or arcs of circles are always measured, and angles also, or the opening between the two straight lines called radii, reaching from the centre to the circumference of any circle, whether the circle is actually drawn or only imagined to be drawn. For instance, the angle 90° means the opening at the centre of the circle between two lines drawn to the two ends of a quadrant of the circumference ; and 90° is called a *right angle*, and lines at right angles are also said to be *perpendicular* to each other ; for ' perpendicular ' in mathematics does not always mean upright, though an upright stick is of course perpendicular to a piece of level ground or the surface of water, which is always level. If you stick a pencil with its flat end upon a globe, it is perpendicular to the surface of the globe, or upright.

All the meridians of the earth then (treating them as circles, notwithstanding their slight ellipticity) are divided into 360°, and those degrees are measured from the equator toward each pole, and every such degree is called a *degree of latitude*, and therefore each pole is at latitude 90°. Again, the distance of any meridian, measured in degrees on the equator and on all the small circles of latitude as well, from any other meridian which is used as a standard or 0 (called zero), is its *longitude*, and is the longitude of all the places on it. A degree of longitude contains 69·17 miles at the equator; but as you go further north or south, the meridians come closer together, and so a degree of longitude measures fewer miles the further you get from the equator. In England a degree of longitude measured east or west from the meridian of Greenwich Observatory, which is our zero, is about 43 miles, whereas a degree of latitude measured *on* any meridian, is about 69 miles everywhere—that is, 69·4 in high latitudes, and 68·8 near the equator, for the reason which you will see presently.

Long before anybody attempted to measure the difference between the equatorial and polar diameters, Sir Isaac Newton, who was born on Christmas Day, 1642, and died in 1727, calculated what it ought to be;

though calculation will not make it quite right, from our ignorance of the density of different parts of the earth. It was probably once all fluid, like the lava from volcanoes, with all the water hanging over it as steam ; and even now it gets 1° hotter for every 90 feet down a mine, and water is hotter as it comes from greater depths.* It would then take the shape of a globe, like a drop of rain, or melted lead in making shot, because the mutual attractions of the particles balance themselves in that shape only. When it began to spin it would swell at the equator, and shrink at the poles, as a large elastic hoop will do if you spin it quickly round its diameter. Newton calculated how much extra weight laid on the equator would balance the loss of weight or gravitation to the centre there, by reason of the centrifugal force arising from the spinning ; which increases as the square of the velocity of rotation ; † *i.e.*, it would be four times as great if the earth turned twice as fast, and if it turned round in an hour and 25 min. people could not stand at the equator, but would be thrown off. Besides that, he had to calculate how much the attraction to the centre is altered by the alteration of the shape from a sphere to a spheroid, and the result is compounded of those two calculations (see p. 37).

If you wish to know how the circumference and the polar and equatorial diameters of the earth are measured, it is done thus. As the earth turns once round or 360° in 24 hours, two places whose meridians are

---

* Appendix, Note II.

† Centrifugal force also increases as the diameter, but so does the attraction of a globe upon its surface, which counteracts it (see p. 29).

crossed by the same star at an interval of 4 minutes,
are 1° of longitude apart; and the equatorial circumference (treated as a circle) is 360 times the measured
distance of two such places on the equator. But the
rotary motion of the earth is of no use for measuring
latitude; and besides that, meridians of the earth are
not circles but ellipses, and the plumb line, or a line
perpendicular to the surface of water, does not point
quite to the earth's centre, in consequence of the spheroidicity, though we popularly say it does, except at the
equator and the poles. Places are 1° apart when their
plumb lines make an angle of 1° with each other, which
can be measured by the stars, as they are so far off that
they may be used as fixed points in the great sphere of
the heavens with its centre at the centre of the earth.
It is found by measuring in this way that 1° of latitude
in Sweden is ⅔ mile longer than at the equator, and
100 yards longer in Scotland than in the south of
England. And thus, by taking a few degrees at
different latitudes, a whole meridian, or section of the
earth through the poles, can be made up and measured
both in shape and size, and the difference of the equatorial and polar diameters ascertained. A nautical or
geometrical mile or *knot* is 1′ of longitude at the equator, or about one-sixth longer than a common mile.

MAPS.

One consequence of the earth being round is that no
map of any large part of it can be correct. You cannot make a large piece of paper lie close upon a globe
without crumpling the edges. Therefore if the middle

of a country is drawn on the map as it would be on the globe, the outsides would be drawn too large, and *vice versâ;* and the larger the country is, the more some parts of it must be enlarged beyond others, or distorted. Maps are made on various plans, some distorting the country in one way, and some in another. The common 'map of the world,' in two flat circles, makes the equator only twice as long as the diameter of the earth, instead of $3\frac{1}{7}$ as long. And each of those two circles, which stand for the *hemispheres*, or half the surface of the globe, only show half as much surface as a hemisphere of that diameter really has.

That mode of 'projecting' a hemisphere or any part of it on a plane is called the *orthographic*, because it shows the surface as it would be seen *straight* by parallel lines of sight from an infinite distance. It represents the middle of the country tolerably right, but the outsides are crowded, and very much so toward the edges of the hemisphere.

If you suppose the globe transparent and the eye at the *antipodes*, or the opposite end of a diameter to the country looked at, it is seen projected *stereographically* on any plane at right angles to that diameter, such as a plane touching the sphere at the other end of the diameter, or a glass plate parallel to that and nearer to the eye. This is in fact a perspective view of the country from the middle of its antipodes; for perspective only means transparent, and the glass plate theory is the foundation of all perspective drawing. But the country must be reversed for the map, as it is seen from the inside instead of the outside of the globe. This

method has the advantage of preserving the shape of
every part, but the middle parts are now more crowded
than the outsides, though not so much as the converse
in the orthographic projection.

The *gnomonic* projection has the eye at the centre of
the transparent globe instead of the surface, and is seen
upon a plane touching it, or on any transparent plate
parallel to that plane. This too is sufficiently accurate
for countries near the point of contact, but at any considerable distance from it the outsides are very much
exaggerated, as you may easily see if you draw a circle
touching a straight line, and divide it into some equal
parts and draw lines from the centre through those divisions to that line, which represents the tangent plane.
It has the advantage that all great circles through the
point of contact are opened out into straight lines crossing there, and all small circles parallel to the plane are
projected into circles with that point for their common
centre. This is frequently used for star maps, the plane
of projection touching at the pole, so that all meridians
are straight lines crossing there ; but circles of declination (which correspond to parallels of latitude on the
terrestrial globe) are not equidistant, as they are in—

The equidistant projection, which has the eye above
the sphere at a height=half the chord of a quadrant,
or ·707 of the radius, looking straight down through
the globe upon a tangent plane at the opposite pole.
This makes the latitude or declination circles for a
considerable distance nearly equidistant, and the meridians straight lines as before; and so this is the best
kind of map for the regions round the poles.

In *Mercator's projection*, which is a favorite one for maps, the globe is supposed to be stretched out on the inside of a cylinder which touches it all round the equator, and the cylinder is then cut and opened out flat or 'developed.' But besides that, since *parallels* of latitude in England would be stretched wider to fit the cylinder in about the proportion of 43 to 69 (p. 17), therefore *degrees* of latitude or the length of pieces of the meridian are drawn in such maps wider than those near the equator in the same proportion, in order to keep the true proportions between the length and breadth of each division of the map; and so the dimensions increase rapidly towards the poles. For this reason it is unfit for maps near the poles; and the maps of countries of high latitude must be made on a different scale from those near the equator, or rather, as if they had been developed from a different globe, in order to get them on anything like the same scale.

There is yet another, very convenient for some purposes, called the *conical* projection. Suppose you want to map a country in the latitude of England. A hollow cone is supposed to be dropped over the globe, of such an angle that it will touch it all round at latitude 52°, and therefore the top of the cone will be vertically over the north pole. Then the country is drawn as it would appear on the inside of this cone to an eye at the centre of the earth, and the cone is 'developed.' Consequently, the meridians are all straight lines converging toward a point which was the top of the cone, and the parallels of latitude are nearly equidistant, and in fact are drawn quite so for convenience. In this there is scarcely any distortion for a moderate breadth of country from north to south, or a *zone* between two parallels of latitude near to the circle of contact with the cone. ⁀This also is used for star maps, and so, indeed, are all the projections, except Mercator's and the orthographic, and they all have their advocates.*

Sir J. Herschel remarks that London is very nearly the centre of that hemisphere of the globe which contains more dry land than a hemisphere described round any other place as its pole. Those who have read a little of Greek history know that a certain place, Delphi, was called the *navel* of the world, being then supposed to be the middle. The real one, you see, is not in Greece, but England. In order to see this, take a terrestrial globe, and elevate the north pole 51½° above the north side of the wooden horizon, and bring London up to the brass meridian : then all above the horizon is the hemisphere with London for its pole or highest point, and it includes all Europe and Africa, and all Asia except a few promontories, and all North America and most of South, leaving only the rest of it, and Australia and some islands, to the other hemisphere.

The following proportions of land and water over the globe, and the north and south hemispheres, and the five continents, with their islands, have been ascertained by weighing paper patterns of them taken from

---

* See Enc. Brit. ' Geography (mathematical),' and a chapter on this subject in Proctor's ' Handbook of the Stars,' a book chiefly of tables. There are other projections of the sphere, but these are the principal.

a globe. All the water 145 million square miles, and all the land 52; water north of equator 59, land 39; water south of equator 86, land 13: land in Asia 18, Africa 12, North America 8, South 7, Europe 3½, and Australia 3½.

The earth's surface is four times the area of one of its great circles, or 3·1416 times that of a square surrounding it; only that has to be reduced about a 400th for the ellipticity of a 298th; and the result is 197 million square miles. The surface of any zone, or band round the earth between two parallels of latitude, is proportionate to its thickness, neglecting the ellipticity. Therefore the surface of each hemisphere is divided equally at 30° of latitude; for a cut through there cuts the axis half way between the pole and the equator.

The solid content of a globe is ·5236, or a little more than half of the surrounding cube. And the bulk (but not the surface) of any spheroid is to that of the sphere which touches it all round as their different axes. Therefore the earth is a 298th less than the sphere which would contain it. But a sphere *as large as* the earth would only have a diameter of 7917·2 miles; for the cube of that = the polar axis of 7899·5 × the square of the mean equatorial diameter 7926, if these figures are right (see p. 169 note). Therefore the earth contains 259,845 million cubic miles.

WEIGHING THE EARTH, AND LAW OF ATTRACTION.

But a far more difficult thing has been done with the earth than measuring, and that is weighing it; if it can be said to have been done yet with certainty. Sir Isaac Newton, by what Sir John Herschel well calls 'one of his astonishing divinations,' hit upon the very weight for the earth, which is nearly the average of all the modern experiments and calculations, viz., that it is about $5\frac{1}{2}$ times as heavy as if it were all made of water, or half as heavy as lead : which is expressed by saying that its average density or specific gravity is $5\frac{1}{2}$, that of water being always taken as 1, except in speaking of airs or gases. The earth is on the whole about twice as heavy as if it were all made of the hardest and heaviest stones ; but beyond that, we know nothing of its composition. Whatever the inside is made of, it must be squeezed together with tremendous pressure by the weight of all above it, and so it may be a great deal denser than any materials of the same kind near the surface.

I can only give you a very general account of the different contrivances for finding the weight of the earth. One is, by seeing how much the plumb bob is pulled aside by the attraction of a mountain near it, and comparing that, by calculations known to philosophers, with the effect of the attraction of the whole earth ; which can be most accurately measured by the time of vibration of a pendulum, as you will see in a later part of this book, where we shall weigh the sun. This is called the Schehallien experiment, because it was first made at the mountain of that name in Scotland.

The nature of it is this. If two plumb lines are hung 100 feet apart, they make an angle of 1″with

2

each other, because each is pointing, we may say, to
the centre of the earth; and therefore at 6000 feet
apart, or rather more than a mile (5280 feet), the
plumb lines are inclined 1′ to each other. But if there
is a great mass of mountain rising between them, that
will attract each of the plumb bobs, and draw them
nearer together, because all matter attracts all other
matter, as you will hear more fully as we go along.
The mass of the mountain can be calculated with tol-
erable certainty from its size, and weighing specimens
of the rocks which it is composed of; for you must
know that *mass* in mathematics means not merely size,
but size and density together; in fact, what we commonly call weight, only there are reasons for keeping
the two words distinct in some mathematical calculations. It is possible to calculate how much the mountain ought to draw the plumb bobs aside, and make
them converge more than 1′, if the whole earth were of
the same density as the mountain. But in fact the
mountain never does attract them so much as it ought
on that supposition. Therefore that supposition of the
mountain being as dense as the average density of the
earth is wrong; and they can calculate how much
wrong, or how much the average density of the whole
earth exceeds the ascertained density of the mountain,
and they find that the earth must be on the whole
about twice as dense or heavy as if it were all made
of the same rocks as the Schehallien, and about $5\frac{1}{2}$
times as heavy as water.

Another mode of weighing the earth is to see how
much faster a pendulum goes at the bottom of a deep

mine than at the top. If the earth were all the way through of the same density as the rocks near the surface, the pendulum would go slower, as I will explain presently. But it does go faster; and that proves that the earth gets much denser. The only experiment of that kind that has yet been made however, by the present Astronomer Royal in 1854, has given a density so much beyond all the other methods, that very little weight can be given to it; especially as still later experiments of the Schehallien kind, made by Sir H. James, the superintendent of the Ordnance survey, at Arthur's Seat, near Edinburgh, have given a density rather below than above the old amount of $5\frac{1}{2}$ times that of water.

The mine experiment has been altogether unfortunate: once before, the instruments got on fire, and another time a great piece of rock slipped and stopped the operations. I will now explain the reason of what I said about the pendulum losing or gaining according to the density of the earth. We can hardly stir a step in 'physical astronomy,' or the astronomy of causes and effects, without having the law of gravitation forced upon us, and therefore you had better learn at once what it is.

Law of Gravitation.—Newton assumed, what all the results prove to be true, that every atom of matter in the universe attracts every other with a force which increases as the square of the distance decreases: *i.e.*, attraction is 3 × 3 or 9 times as strong at a third of the distance, and so on. Also a body composed of 3 atoms must attract three times as strongly as one. Therefore

the law of gravitation is, that the attraction of one
body A upon another B varies as the mass, or quantity
of matter, or what is commonly called weight, of A,
and inversely as, or as 1 divided by, the square of the
distance between each atom of A and of B, or between their centres of attraction. What is called ' the
attraction of A upon B ' does not depend at all upon
the mass of B, or the reciprocal attraction of B upon
A, but means the *accelerating force* of A upon B;
which is measured by the velocity with which A's attraction makes B move, without any reference to A's
own motion. No alteration of the mass of B makes
any difference in its motion under A's attraction, so
long as A itself is free to move. For although doubling B doubles the attraction between them, or their
relative velocity of approach, or the force with which
they would compress a spring separating them, yet B's
inertia or resistance to motion will be doubled also, and
therefore its *absolute* velocity will remain the same as
before; but A's absolute velocity, and the accelerating
force of B upon A, will be doubled by doubling B's
mass. And so it is true, though it looks like a paradox,
that the earth's attraction moves the sun just as much
as it would move a pea at the same distance. The attraction of the earth, or of the sun, is measured by the
velocity with which things fall or move toward them;
for it is only the resistance of the air that makes feathers or grains of sand fall slower than a lump of lead.

For the next step you must accept the following
things as proved by mathematics : (1) a solid globe of
uniform density attracts everything anywhere outside

it, as if its whole mass were condensed into its centre:
(2) the same is true if the globe is made up of a set of
shells or coats, like an onion, each shell having a different density from the others, provided only each shell
has the same density throughout: (3) a spherical or a
spheroidal shell exerts no attraction in one direction
more than another upon a body inside it; or the body
would float in the air anywhere indifferently inside of
such a shell: (4), which follows from the first, the attraction of two globes of equal density on bodies at
their surface is simply in the proportion of their diameters or radii; for their attractions are directly as the
masses, which are as the cubes of the radii (p. 24); but
the attractions at the surface are also inversely as the
squares of the radii; and dividing the cubes by the
squares, you have the result that the attractions of the
globes at their surface are as their radii.

Then if the earth were of the same density throughout, the attraction on things at the bottom of a mine,
say a mile deep, would evidently be less than it was at
the surface, exactly in proportion to its depth; because
the shell a mile thick all round the earth goes for
nothing when you have got inside it by going down the
mine, and the attraction becomes that of an earth of a
mile less radius. Attraction of the earth near its surface is only another name for gravity, and it is gravity
which makes a pendulum swing, and the weaker gravity is the slower it will swing.

But if this shell a mile thick is of very light stuff,
say for simplicity of calculation, of no weight at all, the
result of going down the mine will be very different;

for at the bottom of the mine there will be now the whole mass or weight of the earth attracting as before, and the pendulum will be brought a mile nearer to the centre, from which all the attraction has to be measured, by our first and second rules.   Therefore the attraction will be greater in the proportion of the square of what you may call 4000 miles to the square of 3999, or will be one 2000th more than at the surface.   And as it is also a mathematical fact that the quickness (the converse of 'time') of a pendulum increases in proportion to the *square root* of the force of gravity, it comes back to this, that the pendulum would gain one second in 4000, or nearly a minute in three days, down the mine of such an earth, instead of losing as in the earth of uniform density.   Between these two extremes, of the outside of the earth being quite as dense as the inside, and being of no density at all, there is of course some medium condition in which the force of gravity down the mine would be exactly the same as at the surface; and there is some other condition between that medium and that of no outside density, in which the pendulum would gain, not a minute in 66 hours, but a second in ten hours; which was the result of the experiments in the Harton colliery.   But the actual density of the shell all round the globe has to be estimated before the mean density of the earth can be calculated from it and the observed gain of the pendulum; and it is pretty clear that that has not been done accurately yet.

The most remarkable of all the earth-weighing experiments is that which goes by the name of Mr. Cavendish, who first performed it in the year 1798. It has since been done again many times over, with every possible provision for accuracy, by the late Francis Baily and by Dr. Reich abroad. Sir J. Herschel warns us that some of the professed explanations of it are radically wrong; and I doubt if anything beyond the following general description can be made intelligible to ordinary readers. This is founded on Cavendish's own account in the Philosophical Transactions. The account of Baily's experiments fills the whole of the 14th volume of the Astronomical Society's Memoirs.

It is a contrivance for weighing the earth against a globe of lead of 6 inches radius. And that is done by comparing the known effects of the earth's attraction on pendulums at 4000 miles from its centre, with the experimental attraction of two such globes, at 9 inches from their centres, on two balls balanced on a long rod hung from its middle by a wire, which gently resists being twisted. When the globes are brought sideways near the balls they move them a little against the resistance of the wire, and make them oscillate very slowly about their new position ; which was 14′, or one 250th of the length of each radius, from the old position ; and each vibration took about 420 seconds, with the wire generally used. For simplicity let each radius or half of the rod, with its ball, be as long as would beat seconds if hung as a pendulum, under the earth's attraction. Then it may be proved that the earth's density is to that of lead, as $250 \times 36 \times$ the square of 420, is to $81 \times$ the earth's diameter in feet, that of the globes being 1 foot.* If a stiffer wire is used the balls move through a less angle, but the square of the time of vibration decreases in the same proportion, and so the result is the same; and all the performers of the experiment have agreed in finding the earth's density between 5·44 and 5·67 times that of water. The Schehallien experiments, and another at Mount Cenis, made it rather under 5, the Edinburgh one 5·3, and Mr. Airy's mine experiments 6·57. But whatever the absolute weight of the earth may be, the proportion between it and the sun and moon and planets will remain the same, as they are all calculated from it. So the earth's diameter is the standard by which all the solar system has to be measured, as you will see hereafter. A mistake in the received size of the earth stopped Newton's belief in his own discoveries for some years, until it was remeasured and corrected.

MOTIONS OF THE EARTH.

As soon as it was discovered that the earth is a globe, it could not require much philosophy to conclude that day and night and the visible motion of the stars were caused by its rotation, rather than by that vast number of heavenly bodies all revolving together round it, while they kept their own distances from each other as if they were fixed in a frame.

The two following direct proofs of the earth's rotation were invented a few years ago by M. Foucault.

---

* A fuller account of this may be found in Mr. Airy's Lectures (lately republished with the title of *Popular Astronomy*); but even that assumes some mathematical results, as this does. See also p. 233.

If a heavy ball is hung by a long string from the ceiling and set and kept swinging, taking care to make it oscillate in one plane and not revolve, it will be seen after some time to be swinging across the floor in a different direction from that in which it started. The reason is that the floor has revolved under it with the rotation of the earth. If such a pendulum were swung at the north or south pole, the floor would revolve under it in 24 hours; at the equator it would not revolve at all; and at intermediate places, such as England, it revolves slower than at the poles, but still enough to be visible in an hour or so. The same thing may be shown in the machine called the *gyroscope*, where a heavy disc or wheel, turning on pivots set in a ring which itself turns on other pivots at right angles to the disc pivots (like the gimbals of a ship compass), will keep spinning in the same plane and with its axis pointing to the same star, while the frame which carries the ring moves round it with the rotation of the earth. This is the more complete experiment, and may be performed anywhere. For if the wheel is spun in any latitude, with its axis in any direction at right angles to the earth's axis, the force of rotation of the wheel will keep the plane of the ring directed to the same stars, while the outer frame turns round it on the other pivots, which will be parallel to the earth's axis.

From that date of 600 B.C. to about 1500 A.D. we hear of only one suggestion that the sun does not go round the earth, but the earth round it. Archimedes tells us in a book of his own, that another astronomer,

2*

Aristarchus, about 280 B.C., held the opinion that the earth goes round the sun in a circle, and that the size of that circle is quite insignificant compared with the distance of the stars. Unfortunately for the credit of Archimedes, he entirely disbelieved it; as we shall see that much more modern astronomers have discredited other people's discoveries which were equally correct. It is true that nothing of the kind appears in the only small book of Aristarchus himself which has come down to us, 'On the sizes and distances of the sun and moon;' but there is nothing there contradictory to it, and he may have discovered the motion of the earth after he had written his book on its distance from the sun.* One can hardly suppose that a man like Archimedes would take the trouble to combat the opinion of another mathematician on such an important question without knowing that he held it. It is quite certain, however, that Aristarchus knew the rotation of the earth, and that it is the earth's shadow that eclipses the moon; and that he had a tolerably correct idea how much further off the sun is than the moon, though he did not know the real size, and therefore the real distance, of either of them. Some writers have given Pythagoras credit for teaching that the earth goes round the sun, about 500 B.C., but apparently with no good reason.

But if Aristarchus satisfied himself that the earth moves round the sun, the world itself and even the best astronomers were not moved into believing it for

---

* I take these statements from the 'Lives' of these two philosophers, and quotations therein: I do not profess to have read their books.

nearly 2000 years more; that is until the time of Copernicus, a German priest, who was born in 1472, a few years before a still more celebrated reformer of old opinions, Luther. And for a good while the astronomers and everybody else refused to believe him any more than Aristarchus, and invented all sorts of ingenious theories to account for the visible motions of the planets. But the more those motions were examined, the more impossible it was found to reconcile them with any theory except that of Copernicus, which treated the earth and the planets as all revolving round the sun. Then came Galileo, a still greater astronomer of Florence, who died the year Newton was born, and invented the telescope, which gave him the power of examining their motions still more accurately; and he found other proofs of Copernicus's theory, and (as is well known) was imprisoned for three years by the Roman Inquisition for publishing them. Afterward, about the year 1680, Newton made the far greater discovery of the reason why the earth and the planets and the moon all move, and must move for ever as they do, founded on that law of gravitation which I have already stated. I will explain afterward how that law affects them; for the present we will go on with their motions as they are.

The earth then goes round the sun, so as to see the same stars again in a line with him, in a year of about 365¼ days; a day meaning the time of one rotation of the earth on its own axis from noon to noon, or the average time of the sun's twice passing the same meridian. But we shall have to consider the length of the year more exactly afterward, and also of different kinds of days which are dealt with in astronomy.

If the world were habitable all round, and not divided by the Pacific Ocean, which is opposite to Europe, it would be impossible to avoid a sudden break of time somewhere, which would make the same day March 20 on one side of the boundary and March 21 on the other side. For the days begin and end later as you go west, and earlier as you go east; and if a man could sail round the earth westward he would find that he had lost a day in his reckoning by the sun when he came home, and that he had gained one by sailing round the globe eastward, as the earth turns from west to east.

The earth's mean distance from the sun (that is, the average between the greatest and least distances) is 91,404,000 miles, according to the latest calculations,* and therefore the whole length of the earth's path or *orbit* is 574,310,000 miles. If you work it out you will find that that comes to a rate of travelling through space of 65,518 miles an hour, or 18·2 miles in a second, —or 80 times as fast as sound, or an ordinary cannon-ball, goes through the air.

The reason why we neither feel moving with this enormous velocity, nor are thrown off by the centrifugal force of the earth's rotation, is that both motions are steady, and motion without shaking is as easy as rest. We carry the air with us, as in a railway carriage, and therefore feel no wind. Also the attraction of the earth is 289 times greater than the centri-

---

* Appendix, Note III.

fugal force even at the equator where it is greatest.
Nevertheless things *are* lighter there than in high latitudes. A spring balance which weighs rightly at the
poles would mark 289 pounds as weighing 288 at the
equator, from the centrifugal force alone (pp. 16, 241).

Things weigh less at the equator from another cause
besides. An oblate spheroid attracts less there than
at its poles (but see p. 197). If the earth's density
were uniform, the difference would be only a fifth of
the ellipticity or $\frac{1}{1490}$; but the inside is much denser
than the outside (p. 26). If the outside had no density
at all, the equatorial attraction would be to the polar
inversely as the squares of the two radii, or a 149th
less; for when quantities differ by a small fraction,
their squares differ by twice that fraction. In fact
the equatorial attraction from this cause is a 590th less
than the polar. Therefore taking both losses together,
194 pounds at the poles will only weigh 193 in the
same spring balance at the equator ; and between
London and the equator 1000 pounds lose about 3, and
a clock pendulum loses $2\frac{1}{4}$ minutes a day.

Another objection may occur to you, as it did to
those who imprisoned Galileo, that in several places in
the Psalms it is said that 'the earth shall never move,'
and so forth. But that has been answered by the late
Dr. McCaul, who showed that the Hebrew word which
is translated ' move,' really means to shake or totter ;[*]
and so those passages of the Bible, instead of contra-
dicting the truth, were only waiting to confirm it, as
soon as the truth itself was discovered by the advance

---

[*] Aids to Faith, p. 219.

of science; for the stability of the universe against
shocks and permanent disturbances is now proved to
be a consequence of the law of gravitation.   The same
may be said of the famous passage, ' Let there be light,'
which people used to admire for its poetic grandeur,
but they had no idea till lately that no other words
would have been equally correct; for it is now ascertained that light is not a thing to be created, like
water, but rather a state of things, like fire or noise.

ELLIPTIC ORBIT OF THE EARTH.

Hitherto I have spoken of the earth going round the
sun in a circle, as it does very nearly, but not quite;
for the earth's orbit is an ellipse, and not a circle.   It
is remarkable that another ancient astronomer, Hipparchus, about 150 B.C., found out that the (apparent)
orbit of the sun round the earth was not quite a circle,
though he stopped short of the greater step which
Aristarchus had probably made for himself 100 years
before, and missed observing that the apparent orbit
of the sun round the earth is the same as the real
orbit of the earth round the sun.   This figure of the
ellipse is so important in astronomy, that you had
better learn at once what it is; for every oval or figure
like a flattened circle is not an ellipse.   Ovals for
picture frames are often made out of four pieces of
circles, two of a large one and two of a small put
together; but no pieces of any circles will make a real
ellipse.   The simplest way to make one is to take a
piece of thin string with a loop at each end; stick a
pin through each loop into the table through a sheet

of paper, leaving the string quite loose between them; put a pencil in to stretch the string out, and run it along, always keeping the string tight: then the pencil will describe an ellipse.

Here is a figure of an ellipse with the circle containing it. SPH is the string: SH being the places of the pins, each of which is a *focus* of the ellipse: C is the centre of both the ellipse and the circle; ACD is the *axis major* or the greatest diameter, which evidently = the length of the string or SP+HP, or twice SB. BCF is the *axis minor*. The nearer together the foci are, the more the ellipse approaches a circle, or the less eccentric it is: the proportion of CS to CA (not CS alone) is called the *eccentricity*, which is therefore expressed by a fraction, either a vulgar fraction or a decimal, as may be most convenient; and you see it is much greater than the ellipticity $\frac{EB}{AC}$ (p. 15): when the ellipse is very nearly a circle the ellipticity may be called half the square of the eccentricity. We speak of the eccentricity of the planets' orbits, because we are not concerned with their centre but their focus, where the sun always is. On the other hand we have nothing to do with the focus of a meridian of the earth, and so we speak of its ellipticity. SB or AC is also the ' mean distance.

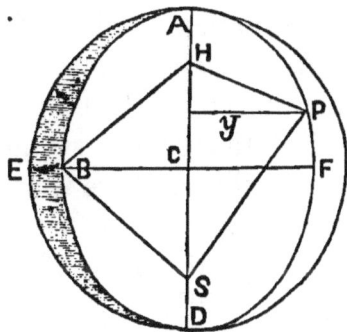

If you want to find the minor axis of a planet's orbit from the major axis and eccentricity (which are always the things given) you may do it by this rule: the square of the linear eccentricity, *i.e.*, SC^2=the sum of the semi-axes × their difference; the reason of which will be evident to any one with a little knowledge of mathematics. Taking SP alone, which is called the *radius vector*, it evidently varies more or faster the more eccentric the ellipse is.

But though the string method is the best for drawing an ellipse, there is another definition of it which it is important to understand. An ellipse is the oblique or perspective view of a circle. For if the circle is turned a little on its diameter, it will cover from your eye the elliptic space ABDF, and all the lines at right angles to that diameter, such as CF and *y* (which letter is always used in mathematical books for those lines, called *ordinates*), will be less, in proportion to the ellipticity EB, considering AC invariable. But the centre of the perspective ellipse does not coincide with that of the circle, or fall upon the line of sight from the eye to the centre of the circle, except when it is seen so far off that all the lines of sight may be considered parallel, as in looking at any of the heavenly bodies.

The ellipse is the shape of the orbit of all the planets, and of the moons round any planets which have any. The linear eccentricity of the earth's orbit is ·0168, or one 60th, of its semi-axis major or mean distance of 91,404,000 miles: and therefore the earth is about three million miles nearer the sun at one time than another. You may very likely think this is the cause of the difference between winter and summer; but it is no such thing. On the contrary, the earth happens to be nearest to the sun in the middle of our winter—the 1st of January, though it has not always been so, and will cease to be so again. For the whole ellipse turns round, going forward, or in the same direction as the earth itself moves, $11''\cdot8$ a year, or 1° in 308 years; and at the same time the equinoctial points, on which the times of all the seasons depend, as you will see presently, go backward at the rate of $50''\cdot1$ a year, or completely round in 25,868 years. And as one goes one way and the other the other way, it is the same as if the places of *perihelion* and *aphelion*, or nearest and furthest distances from the sun, went forward at the rate of $61''\cdot9$ a year, or completely round in 20,984 years, relatively to the equinoxes, from which all celestial measures are taken, though the time of absolute or sidereal revolution of the perihelion is 109,830 years.

The time the earth takes to return to perihelion is called the *anomalistic* year, because the distance of a planet from perihelion, or of the moon from perigee or point of nearest approach to the earth, is called its *true anomaly;* and the distance it would have gone in the same time if it moved uniformly, or in a circle instead of an ellipse, is its mean anomaly; and their difference is called the *equation of the centre:* all these being measured by the angles described by the radius vector round the sun—or earth in the case of the moon. The anomalistic year is 25m. longer than the equinoc-

tial year, in consequence of the advance of the perihelion.   But this is only a fact, and not a period used for
calculation.

You see from the above rate of advance of the perihelion that the earth was nearest to the sun in summer,
and furthest off in winter of the northern hemisphere,
about 3600 years before the creation of Adam.   And
then it was much hotter in summer and colder in winter; Sir John Herschel calculates, no less than 23°, a
very serious difference indeed, and other calculations
make it more.   To some extent that is so now in the
southern hemisphere, and the summer is much hotter
in Australia and South Africa than equally far north
of the equator; but it would be much worse when all
the land of the north hemisphere was exposed to 23°
more heat and cold, whereas now it is chiefly the sea
that receives it, on which heat makes less impression.

Moreover it is calculated that the eccentricity of the
earth's orbit was formerly much greater than it is now,
in the following proportions:

Present eccentricity .   .   .   .   ·0168
  50,000 years ago. .   .   .   .   ·0131
 100,000   "   " .   .   .   .   ·0473
 150,000   "   " .   .   .   .   ·0332
 200,000   "   " .   .   .   .   ·0569
 210,000 maximum.   .   .   .   .   ·0575
 250,000   "   " .   .   .   .   ·0258
 300,000   "   " .   .   .   .   ·0424
 350,000   "   " .   .   .   .   ·0195
 400,000 same as now. .   .   .   .   ·0170

and then the eccentricity increased again.   Whenever
in that long period of great eccentricity, from 80,000

to 300,000 B.C., the earth was at aphelion in the northern winter it was much colder than it is now. Let us see how much. When the eccentricity was ·0575, the sun's greatest distance was nearly 97 million miles, against our present nearly 90 in winter. Our average winter temperature is called 39°. But this is only 39° above an arbitrary zero, which is of no use for measuring the power of the sun. We must reckon from the absolute zero, or the heat of no sun at all, which is estimated from various experiments to be 490° below our zero : so that our winter heat is really 529° on the absolute scale, or the sun raises the temperature so much above what it would be if there were no sun. And the heat is inversely as the square of the sun's distance. Therefore the average heat of the winter of Europe about 210,000 years ago was $\dfrac{8100 \times 529°}{9409} =$ 455°, or 74° below the present heat, whenever the northern winter happened at aphelion.* And as the aphelion revolves relatively to the equinoxes in 21,000 years, there must have been ten of these northern winters in aphelion during the period of great eccentricity : not all so intense as the one 210,000 years ago, but all far beyond our present cold; and nearer the pole of course it was colder still.

Every such time was probably a *glacial epoch*, as it is called by geologists, when all Europe was covered with ice, which the heat of summer had not time to

* See Mr. James Croll's paper in the Philosophical Journal of October, 1865, and a much fuller one in February, 1867, confirming these conclusions by many more reasons ; and Tyndall on heat, p. 79.

melt, being also obstructed by the evaporation from the melting snow; and which slid down our valleys like the glaciers in the Alps, and as icebergs slide into the Arctic seas. Moreover it is thought that the weight of that ice was enough to shift the centre of gravity of the earth and keep down most of the land of the northern hemisphere below the level of the sea, as geologists say it has been. Then also the winter was nearly a month longer than the summer, as it is now a week shorter, because the earth moves quickest when it is nearest to the sun; though the heat received by the whole earth is the same in each 180° of its revolution, the longer time of the distant half of the orbit making up for its greater distance: but that is by no means the case for each hemisphere separately. The heat received in a given time by any planet is inversely as the area of the orbit, or as 3·1416 × the product of the semi-axes. But the axis major is practically constant, and the gradual increase of our semi-axis minor, which is now only 12,800 miles less than the major, is insignificant for this purpose, though not for another (see p. 141).

The Seasons.—The real cause of summer and winter is that the earth's axis does not stand upright in her orbit round the sun, which is called the *ecliptic*, but one pole always leans 23° 28′ toward what we call the north of the heavens or fixed stars, and the other pole leans as much to the south. Consequently, when the earth is on the south side of the sun, the north pole, and the northern hemisphere generally, are turned toward the sun, and the south pole away from him, and it is summer in the north and winter in the south.

Six months after, the earth having gone half round the sun, the north hemisphere is turned away from him, and the south hemisphere then looks toward the sun, and so it is winter in the north and summer in the south.

Summer is not only the time of warmth, but also of longest days; and to explain that we must consider the earth's rotation as well as revolution. You had better take a terrestrial globe and elevate the north pole 23½° above the wooden horizon, which we may take to represent the boundary of light and darkness, assuming the sun to stand right above it; for of course there is no such thing really as 'above and below' in the heavens, and we only use these terms for convenience. Then as you spin the globe round for its daily rotation, you will see that nothing within the *arctic circle*, 23½° from the north pole, ever goes below the horizon or into darkness; that is, the sun never goes below the horizon of the people within that circle in the northern midsummer. At the same time nothing within the *antarctic circle*, 23½° from the south pole, comes into the light at all, and so those people have no daylight in the middle of their winter. The converse of all this evidently takes place at the opposite time of the year. But half way between those times, when the earth is either east or west of the sun, the two poles are equidistant from the sun; and so the light received by the two hemispheres is equal. And to measure the length of days at those times you must lay the poles level with the horizon or boundary of light, and you will see that every part of the globe is

just as long above as below it, or the days and nights
are equal.

Therefore those times are called the *equinoxes*, which
occur on March 21 and September 23. Midsummer
and midwinter are called the *solstices*, because the sun
then stays at the same distance from the equator for a
few days, and the days remain of the same length
before they begin slowly to get shorter or longer again.
From the solstices to the equinoxes the two poles gradually approach to an equal distance from the sun ; and
so the difference of days and nights gradually diminishes to 0, and after the equinoxes increases again.
At the poles they have a winter of six months in which
they never see the sun, and a summer (though a very
cold one) in which the sun never sets. The nearer
you go to the poles the greater is the difference of days
and nights at all times except the equinoxes. Even in
the north of England the difference is visibly greater
than in London and the south.

At the equinoxes the sun appears on the equator,
which means in astronomy not merely a circle round
the earth equidistant from the poles, but the plane of
that circle extended to the heavens. And as the sun
is always in the ecliptic, the equinoxes are the places
where the equator and ecliptic cross each other. For
all these purposes we may properly talk of the earth
as turning on a fixed axis, and the sun moving round
the earth in the ecliptic. For their relative motions
are just the same as if they did so ; and if there were
no other bodies in the universe to measure by, no
human being could ever have found out that the

earth does not stand still with the sun revolving round it.

The equator is necessarily as much inclined to the ecliptic as the axis of the earth is to that line perpendicular to the ecliptic which is called the *axis of the ecliptic*, whether it is in the sun or the earth. Therefore two circles, each 23° 28′ from the equator, are the boundaries of the sun's journey to the north and the south of the equator. They are called the *tropics*, which means the turning places of the sun ; the northern one is the tropic of Cancer, which the sun touches at our midsummer, and the southern is called the tropic of Capricorn, which the sun reaches in our winter and the southern midsummer. The band between the tropics is called the *torrid zone*, the two arctic circles the *frigid zones*, and the spaces between are the *temperate zones :* and they cover respectively ·4, ·08, and ·52 of the earth's surface.

But you may ask why should the torrid zone be always hot, since the sun is nearly 47° away from each tropic when he is at the other, while he is almost directly overhead to parts of each temperate zone when he is at the tropic nearest to it. The reason is that places within the torrid zone get a greater quantity of sunshine nearly or quite direct in the whole year than any places can outside of it ; and the heat is accumulated, or as it were bottled up in the earth, and stays there after the sun has left that place or latitude. The heat received anywhere depends on the directness of the sun's rays, or its apparent verticality overhead ; for a square foot or a square mile of surface evidently catches more or less rays from a fire or the sun according as it faces them directly or is turned obliquely toward them. Now every place within the tropics has the sun directly over it not only once, but twice a year, and has in fact two summers, one as the sun is going from the equator to the tropic and another as he returns over the same latitude toward the equator. In that way a quantity of heat is accumulated which no place beyond the torrid zone can get. Still it is sometimes hotter in the low latitudes of the temperate zones than it is at other times within the torrid zone. More heat is gained in long sunny days than radiates away in their short nights, and the excess accumulates and makes July and August hotter than June and May.

The equinoctial points, where the planes of the equator and ecliptic cross each other, are of great importance in astronomy, because nearly all the celestial measures are reckoned from the point of the vernal equinox, which is called the first point of Aries ♈; the autumnal one being the first point of Libra ♎. Aries and Libra are the names of two clusters of stars or *constellations*, which were imagined by the ancients to represent a ram and a pair of scales; and they are also still kept as the names of two of the divisions of the ecliptic into twelve parts called *signs of the zodiac*, which are these:—Aries ♈, Taurus ♉, Gemini ♊, Cancer ♋ (of which the first point is the summer solstice, from which the northern tropic is named), Leo ♌, Virgo ♍, Libra ♎, Scorpio ♏, Sagittarius ♐, Capricorn ♑ (the winter solstice is at the tropic of Capricorn), Aquarius ♒, and Pisces ♓. But they are now seldom used in astronomical books. About 2200 years ago the sun used to appear entering the constellation Aries when he also entered that first point of the sign Aries, or the equinoctial point. But now the sign Aries has left that constellation, for the reason you will see presently, and is in the constellation Pisces.

Trade Winds.—The heat of the torrid zone and its velocity of rotation produce the *trade winds*, which blow constantly in the same directions in the same latitudes on the great oceans; though not so constantly on land, on account of variations in heat and other causes of disturbance. The heat expands the air and makes it rise from the equatorial regions, and then the denser air from cooler latitudes comes in, and would make a constant north wind (in this hemisphere) if the earth were either stationary or cylindrical. For if the earth had no rotation the air would have no east or west motion; and if it were a cylinder all the air would be carried round with it from west to east with the same velocity, and would be no more felt as a wind than the air you carry with you in a railway carriage, though it moves as fast as a very high wind. But the earth's surface moves 1040 miles an hour at the equator, 900 at latitude 30°, only 520 at latitude 60°, and less still toward the poles, where its velocity from rotation sinks into nothing.

Therefore while the air comes south (in our hemisphere) it is always coming to a place which moves faster eastward than the place it came from; and so that north wind becomes N. E. relatively to the surface

3

of the earth : just as a weathercock would point N. E. in a north wind if you carried one with you running east.   This is the principal trade wind, blowing from N. E. in the northern hemisphere and from S. E. in the southern, up to about latitude 30°.   Near the equator its eastern character is lost, because there is no material increase of velocity in the earth as you get very near the equator.   Also the north and south winds meet there, and make a calm : but the line of greatest heat and calm is a little north of the equator.

The air which rose from the equator must go somewhere, and it goes in an upper current toward the poles, and begins to fall again when it gets cool, to fill up the space left by the air coming to the equator. And as that air from the equator started for the north with an eastward velocity of 1040 miles an hour, and comes down again on latitudes which move much slower, it is felt there as a S.W. wind in this hemisphere, and N.W in the other.   This secondary or *anti-trade* wind prevails from about 30° to 60° latitude at sea, and makes ships sail from North America to England nearly twice as fast as from England to America.

PRECESSION OF THE EQUINOXES AND LENGTH
OF THE YEAR.

I said at p. 35 that the earth goes round the sun, so as to see the same star again in a line with him, or in *conjunction*, in about 365¼ mean solar days.   The exact time is 365·2563 days (omitting further decimals), and that is called a *sidereal* year.   But the important thing for all practical purposes is the year of seasons, called the tropical or equinoctial year; and that is a little shorter than the sidereal for this reason. The equinoctial points ♈ and ♎ recede among the stars. If that imaginary line where the plane of the equator cuts the plane of the ecliptic points this year to any two given stars in the east and west, next year it will point 50″·1 away from them; so that the sun will reach the spring equinox, or cross the equator from south to north, 20⅓ minutes before it comes again into conjunction with the same star as before. Consequently, if we reckoned by sidereal years, the seasons would get sensibly wrong in no very long time. In fact the equinoctial points are known to have moved 30°, which is equivalent to a month, or from the constellation Aries into Pisces, since the early days of astronomical records 2200 years ago. This is called the *precession of the equinoxes*, because it makes them precede their sidereal time; and was discovered by Hipparchus about 150 B.C.

The length of the equinoctial or tropical year is now settled by astronomers—at least by the English ones—to be 365·242216 mean solar days, or 365d. 5h. 48m. 47½s.; and the French measure is practically the same. It is reckoned from the time when a mean sun going at the average speed determined from the longest experience would pass through the mean ♈; for that also has its variations, and the motion of 50″·1 a year is not quite uniform in each year, as you will see presently. But first let us try to realize what kind of motion the earth goes through to produce this effect.

Get a common 'celestial globe,' and set the axis of

rotation upright, and consider the ' wooden horizon '
to represent the ecliptic or plane of the sun's orbit.
Then the axis of the globe will be the axis of the ecliptic, though it is not so on the globe; but we want the
poles of the ecliptic as marked on the globe to represent the poles of the equator or of the earth, for our present purpose.   Now let our new north pole of the earth
lean toward the north side of the room : and you may
consider the sun as going round the earth from right to
left, or west to east (through south).   If there were no
precession the pole would always point the same way,
and the sun would always be among the same stars at
the same seasons.   But the poles of the earth twist
slowly round the poles of the ecliptic westwards, or the
opposite way to the sun, keeping about 23½° from them,
and going quite round in 25,868 years.   Turning our
globe round its upright axis from left to right, you will
see the equinoctial points recede along the wooden
ecliptic : which makes the equinoctial year those 20m.
20s. shorter than the sidereal.

The pole star, or the equinoctial stars, of any epoch
are thus an index to it in the great cycle of precession :
a Draconis, now 25° off the pole, was only 3° 42' when
the Great Pyramid probably was built, about 2170 B.
C., a few centuries after the Flood.   Its entrance passage is 3° 42' inclined to the earth's axis, and looks due
north : therefore that pole star then looked straight
down that long narrow passage exactly at midnight on
the shortest day, and at its lower transit every day.
And at the same time the south meridian was crossed
by the Pleiades in ♉, where the equinoctial point was,

and which are associated with the beginning of the year by traditions and customs all over the world, including the well-known eastern worship of the bull.*

You would not suppose that this precession of the equinoxes was caused by the protuberance of the equator. Newton discovered that it is. The protuberance of the equator is like a heavy ring round the earth. The attraction of the sun acts more strongly on the nearest than on the furthest part of the ring, and would immediately pull the equator straight into the plane of the ecliptic, if the spinning of the earth did not resist it by always carrying off that part of the equator which is nearest to the sun and most attracted for the moment. The result is that the whole plane of the ring is twisted backward, but its inclination to the ecliptic is not altered (see p. 159). The motion is not uniform, but less as the sun approaches the equator; for when he is in the plane of the equator (at the equinoxes) he can exert no force to disturb it.

For the purpose of explaining the cause of precession I have only mentioned yet the sun's attraction; but in fact the moon contributes to it in the same way, and even more: for she also moves nearly in the ecliptic, and attracts the front of the ring or the part which is nearest to her, more than the back parts which are furthest off. Not only is the front of the ring attracted more than the middle, but the middle is attracted more than the back; and so it is on the whole the same as

---

* This is a more complete coincidence than Sir J. Herschel pointed out: see his 'Astronomy,' p. 206; and vols. ii. and iii. of Mr. Piazzi Smyth on the Pyramid.

if the sun and moon pulled the front of the ring or of
the equator up or down toward the plane of the ecliptic, and also pushed the back part down or up. Moreover, as the effect of the sun and moon on precession
is all due to the *difference* of attraction on the near
and far parts of the earth's protuberance, the moon
really does more toward it than the sun, though her
attraction upon the whole earth is very little compared
with the sun's, who is enormously larger and heavier;
but then she is very much nearer; and the difference
between the back and front of the earth is a 30th of
the moon's distance, but hardly a 12,000th of the sun's,
and their differential force is inversely as the cubes of
their distances (see p. 133). The result is, that giving
the moon the benefit of her nearness, and the sun the
benefit of his greater weight, the moon does above
twice as much as the sun in producing the precession;
as we shall see afterward that she also does in producing the tides by the same difference of attraction on
the opposite sides of the earth.

The lunar part of the precession varies like the solar,
according as the moon is near or far from the equinoctial points; and so it may be least when the solar precession is greatest; or they may both be at their maximum or minimum together. The $50''\cdot1$ is the average
or mean precession in a year.

Nutation.—There is yet another irregularity in the
lunar precession. In consequence of the moon being
sometimes a little above, and sometimes below the
ecliptic, she does not pull quite on a level with the
sun, and so produces a sort of nodding of the pole from

its average motion in a circle round the pole of the
ecliptic; and that is called *nutation*. The poles of
the earth, or of the equator, go like a man walking
between the rails round a race course, but with a wavy
motion from one side to the other, instead of walking
along the middle. But the word 'nutation' is used to
comprehend all the variations of precession, both forward and sideways. 9"·2 is the extent of the nutation
on each side of the middle or average course of the
pole in its circle of 23° 28′ radius round the pole of the
ecliptic; and the length of each wave, or rather half
wave, from one crossing of the middle of the course
to another, is 3′ 10″, corresponding to 9·3 years, or
half the time of one revolution of the moon's nodes,
or places where she crosses the ecliptic (p. 101). You
must understand that this 3′ 10″ is not measured
round the pole of the ecliptic as a centre (in which
case it would be 9·3 × 50″·1), but as an arc on the
great sphere of the heavens, with the earth's centre for
its centre, as all the celestial measures are.

The amount of the nutation, or disturbance of the
earth's axis by the moon when acting in a different
direction from the sun, is one of the means used for
calculating her power of attraction, or mass, compared
with the sun's. Another is the comparison of their
effects upon the tides (p. 133). The protuberance of
the earth disturbs the moon in return, as you will see
at p. 197 that Jupiter's oblateness disturbs his moons.

Before we leave precession, I should tell you of a
curious use that has been made of it, toward settling
the question whether the earth is a thin shell of rocks

full of melted lava or other fluid inside, as some persons supposed. The late Mr. Hopkins of Cambridge calculated that the precession would be greater than it is if the earth is not solid for at least 1000 miles deep. For the sun and moon would twist the axis of the earth rather more if the protuberance under the equator were fluid; which may be roughly illustrated thus:—A pendulum with a bob made of a glass globe filled with quicksilver swings rather faster than a pendulum similar and equal in all respects except in the bob being solid. The reason is that the hollow globe does not stick to the mercury and hold it fast, but slides round it as it turns a little in swinging, and so the mercury does not turn with it; whereas the whole of a solid bob has to be turned as well as swung at every vibration, which uses up more of the force of gravity, on which the quickness of the pendulum depends.

Not only do the points of crossing of the equator and ecliptic recede 50″ a year along the ecliptic, but the *obliquity of the ecliptic* itself, or its inclination to the equator, decreases about half a second a year, and will go on decreasing until it has got nearly down to 22° from the present 23½°, when it will increase again. The reason is that the whole ecliptic or plane of the earth's orbit is slowly tilted by the attractions of the other planets. Moreover, as it neither turns on the line of equinoxes, as a hinge or axis through the sun, nor directly across it, it makes the annual precession a little less than it would be otherwise, and also lets it increase a *very* little, so that the tropical year, or year of seasons, is 4 seconds shorter than it was 2000 years ago, though the absolute time of the earth's revolution round the sun has not altered.

We must now consider the different measures of a day more exactly. And first we may remark that the day of astronomers· begins at the noon after the midnight when our common day begins, and has no A. M. or P. M., but simply 24 hours. Thus 11 A. M., 1 January, 1867, was 23 o'clock 31 December, 1866, in astronomical almanacs. But there is also a time called *sidereal*, which is still more different. The sidereal day here begins and ends when the equinoctial point ♈ crosses the meridian of Greenwich. If the sun is on the meridian at the same time (as he is only at the vernal equinox) he will not have quite got there again by the time ♈ is there again, or at the end of that sidereal day ; because the earth has meanwhile moved on a day's journey in her orbit and passed the sun a little, and so has to turn a little more than quite round for Greenwich to face the sun again, since she rotates in the same direction as she revolves round the sun, from west to east, like a small wheel of 8 teeth rolling round a very large one of 2922. For that small wheel would turn $365\frac{1}{4}$ times relatively to the large one in one revolution round it, but $366\frac{1}{4}$ times absolutely, or relatively to one side of the room : which may help to clear some people's ideas about the rotation of the moon (p. 94).

A sidereal day then is practically the time of one absolute revolution of the earth, or the time between

3*

two transits of the same star. For the precession of
the equinoxes makes no sensible difference in a day,
and it is the same thing whether we call a sidereal day
the time between two transits of ♈ or two transits of
the same star. Not only is the daily precession 366
times less than the annual, but the motion of ♈
through 50″·1 only makes a difference of 3⅓ seconds
of time in the time of the clock, or the arrival of ♈
at the meridian at the end of a year, though it makes
20m. 20s. difference in the time of the sun (or earth)
reaching ♈ again, as I said at p. 51. There is necessarily one more sidereal than solar days in a year, or
the sidereal day is one 366th shorter than the solar.
More exactly, a sidereal day is ·9973 of a mean solar
day, or 3 (solar) m. 55·11s. shorter; or a solar day is
1·002738 of a sidereal one; and the hours and minutes
are in the same ratio. The sidereal day never begins
at 12 o'clock, except at noon the 21st of March, and
at midnight of the 23rd of September. It also reckons
by 24 hours, not 12; and its length is invariable (but
see p. 144).

Equation of Time.—We have yet to inquire what a
solar day really is. I call the equinoctial year 365·-
242216 *mean* solar days. A true sun dial solar day
is the time between two transits of the sun over the
meridian. But such days are by no means of equal
length, and so far different that in November the true
solar time is a quarter of an hour before mean time,
which is clock time, and in February a quarter of an
hour behind it. That is the reason why the afternoon
light appears to be so much longer after Christmas

than at the corresponding time before Christmas. The causes of the inequality of solar days are the unequal velocity of the sun in his elliptical (apparent) orbit, and still more, the unequal motion of the sun in the direction parallel to the equator, in consequence of the obliquity of the ecliptic to the equator. Even if he moved quite uniformly, say 1° a day for simplicity, in the ecliptic, you will see if you look at a celestial globe, that meridians through each degree of the ecliptic will not go through each degree of the equator, which is the *hour circle* of the globe. The mean solar day then is the average of all these variable solar days throughout the year. And the thing called the *equation of time* is the difference between the time of day by the clock of mean time and the time by the sun dial. It is the column headed in the almanacs 'Clock before sun,' or 'Clock after sun.' It varies a little from one leap year to another; but a fixed table will do very well for common use with a sun dial, such as is given at p. 12 of the 4th edition of the Rudimentary Treatise on Clocks, with the rule for correcting it from year to year. It is practically the same every fourth year, like the time of sunrise and sunset.

Until a few years ago the clocks in the west of England used to be behind those of London, or Greenwich Mean Time: at Oxford 5 minutes behind, Bristol 10, Edinburgh 13, Exeter 14, Glasgow 17, and so on. The reason of that was that mean solar time, or the average time of noon throughout the year, varies (no less than true solar time) with the longitude of each place; for a place in the east comes to face the sun sooner than one

4*

in the west, as the earth turns from west to east.
When railways brought all places in England in a
manner together, the inconvenience of these separate
local times became intolerable; and the city of Oxford,
which was the last to retain it in the west, has at last
given it up, except at the cathedral. Some places in
the eastern counties still absurdly keep their eastern
time in advance of Greenwich. This difference between
local time and G.M.T. (as it is called) has to be borne
in mind in taking the time from a sun dial or any more
accurate meridian instrument. It has to be added to
' clock after sun ' in west longitude, and subtracted
from ' clock before sun ; ' and the contrary in east.

CHAPTER II.

In considering 'the solar system,' which means the sun and all the planets and their moons or satellites, it is important to get a clear idea of the size and weight of the sun, as it is his attraction that keeps them all in order, moving in their proper times and distances, besides his other equally important business of sending light and heat to them—if they all have inhabitants to use it. The sun's diameter is 845,220 miles, or 106·64 times that of the earth. But that gives you no idea that he is more than a million times as big as the earth; which he is. If you take two bricks and lay them lengthwise, and then put two more alongside, and then put four more on the top of those four, you will have made a thing of the same shape as a brick, but twice as long, twice as broad, and twice as high, and therefore eight times as big altogether; if you treated three bricks in the same way you would make one twenty-seven times as big. And the same is true whatever the shape is; for a body of any shape may be made up of an infinite number of infinitely small cubes: and if each of them is doubled or trebled in every dimension, the body will be doubled or trebled in every dimension, and will therefore be increased eightfold or twenty-sevenfold in bulk. This is expressed by saying that the bulk, or volume, or solid content, and (if the

densities are equal) the mass varies as the cube of the •
diameter.   Therefore the sun is nearly 107 × 107 × 107
or 107³ (as they write the cube of 107) times as large
as the earth ; the actual proportion is 1,217,000, allowing for the earth's spheroidicity as at page 24.

The sun's distance from the earth varies from
89,860,000 to 92,950,000 miles.   The mean between
the greatest and least is called the *mean distance;* and
that is now considered to be 91,404,000 miles, though
you will find it in older books called 95 and 95½ millions, because it was believed to be so until lately.*
You will hear more of that correction as we go on, and
also how the diameter and distance of the sun are both
measured.   It is also 108·2 times the sun's diameter,
and 23,064 times the earth's radius.

As nearly all distances in the solar system are measured by millions, and nobody ever counted a million
of anything, it is worth while to stop a little to understand what it is, by the help of a few specimens.   A
million days are 2730 years ; so there have been not
much more than two million days since the creation
of Adam, and rather more than one million since the
time of Solomon.   A railway train going thirty miles
an hour and never stopping would take nearly four
years to go a million miles, and eleven years to go once
round the sun, and three hundred and sixty years to go
from here to the sun.   If you had a million shillings to
count one by one, and did it as fast as you could for
ten hours a day, it would take a fortnight; and the
million shillings would weigh nearly six tons, or be a

---

* Appendix, Note III.

heavy load for a railway truck. A million is a thousand thousands, and a billion is a million millions, according to the English method of notation, or a thousand millions according to the French method. The French method is usually taught in the schools of the United States.

But instead of describing the size of the sun by these immense figures, you will have a better idea of it by comparing it with something nearer its own size. There is no solid body that we know of at all to be compared to it, and so let us make one out of a distance that we know. A globe large enough to contain the whole *orbit* of the moon, or one with the earth for its centre and the moon rolling on its surface, would only be one-sixth as big as the sun. The sun is quite round, not spheroidal like the earth.

But though the sun is above a million times as big as the earth, he is only 316,560 times as heavy, and is consequently made of much lighter materials. His density or specific gravity is only ·26 of the earth's, or not much greater than water, and rather less than if he were made of coal. You must not suppose that we can measure the sun's density except through his size and his weight, which has to be calculated from his effect upon the earth's motion, as I shall show you toward the end of the book. The density of anything is in direct proportion to its weight and in inverse proportion to its bulk; and as we can calculate that the sun is only a quarter as heavy as he ought to be according to his size in proportion to the earth, we know that his average density is only a quarter of the earth's.

The way to find the density or specific gravity of any-thing heavier than water, which is always taken for the unit, is to weigh it first out of the water and then hanging in water by a string as thin as possible, and the specific gravity is the weight in air divided by the difference of the two weights. It is easy to remember that 36 cubic feet of water weigh a ton ; or a tank six feet square and six feet deep will hold six tons of wa-ter. A cubic foot of water contains 6·23 gallons, and therefore a gallon weighs almost exactly 10 lbs., which it is convenient to remember.

In the table at the end of the book I have given the specific gravities of the sun and planets, assuming the earth's to be 5·5, which is the greatest of them all but Mercury, and very much greater than all but Mercury and Venus. Saturn is as light as deal and the light woods (not cork as Sir J. Herschel says, for that is only ·24) ; and the other three great planets and the sun are about the same as the very heavy woods and coal. Mars and the moon are about as heavy as diamonds and heavier than stones, which are generally 2·7, but much lighter (as the earth is too) than any of the common metals except aluminium, which is only 2·25 or the same as glass. All these densities, remember, are only measured through the earth's, which depends on the experiments described in the first chapter.

I may as well say here, that I have calculated afresh all the quantities in that table which depend upon the measure used by astronomers called the *sun's parallax :* which is the apparent size of half the greatest diameter of the earth as it would be seen from the sun's centre, or half the angle between two lines drawn from the sun to the outsides of the earth at the equator; which manifestly varies inversely as the sun's distance. You will see hereafter how all the dimensions of the solar system (except the moon's) depend on that. I now take the parallax at 8″·943* according to the latest average of observations, instead of the old 8″·57 corresponding to the certainly erroneous sun's distance of 95,383,000 miles. The distance of two bodies always means in astronomy the distance of their centres; and the diameter of a planet means its greatest diameter.

From the diameter and mass of the sun, compared with those of the earth, we can calculate their comparative attractions and the force of gravity at their surfaces. For by the law of gravitation (p. 29) that is directly as the mass and inversely as the square of the distance from the centre. Therefore

$$\frac{\text{gravity on the sun's surface}}{\text{gravity on the earth's surface}} = \frac{316,560}{106\cdot64^2} = 27\cdot83 \ ;$$

or a man on the sun would feel as heavy as if he had nearly 27 others laid upon him, and would be squeezed flat by his own weight.

Sun Spots.—Probably the solid sun is a smaller body inside the luminous shell which we see, and which is now called the *photosphere*, or light-giving sphere of the sun. And the spots which sometimes appear are holes in the photosphere, through which we catch a glimpse of the dark and solid sun. Their depth is estimated at about 30,000 miles; and they have enabled us to measure the time of the sun's absolute or sidereal rotation on his own axis, in 25d. 7h. 48m., and to see

---

* Appendix, Note IV.

that it leans 7° 20′ from a line perpendicular to the ecliptic; and such a line as that, through any body in the solar system, is called the axis of the ecliptic. Therefore the sun's equator is inclined 7° 20′ to the ecliptic : which means that that is the angle between them. The sun's north pole leans most toward the earth on September 13 and his south pole on March 11. Therefore spots move apparently in straight but oblique lines across his face for about the first three weeks of June and December, but at other times in very flat semi-ellipses.

Some sun spots are nearly 50,000 miles wide, and are quite visible without telescopes when the sun is dimmed by a fog. You may wonder how we can say that they are holes in the photosphere rather than dark clouds or patches upon it. If you look at a church with deep windows, the 'splays' or jambs of the windows which directly face you look equally wide on both sides; but the jambs of the windows which you see obliquely appear wider on the far side than the near. And the spots on the sun have sides, less dark than the spots themselves, which appear of equal width while the spot is about the middle of the sun, but wider on the far side and narrower on the near when the spot is away from the middle. These varying sides then are almost certainly the jambs or sloping sides of the spots, which are deep holes in the photosphere. But they are not permanent holes like the volcano craters in the moon; for in some years there are none at all. They have a maximum every 11 years; and they get smaller as they come opposite to

Venus, and to Jupiter in a less degree, but apparently the earth does not affect them :* the reason of all which is yet unknown. The theory which prevailed, of their coinciding with ' magnetic storms,' or spontaneous disturbances of the magnetic needle, seems to be abandoned.* They are confined to two zones from near the sun's equator to about 30° of latitude, answering to the region of our trade winds.

The sun is also estimated to have an atmosphere, reaching 72,000 miles above the photosphere, because ' flames ' or brilliant clouds are seen as high as that when the sun is darkened by the moon in a total eclipse. Moreover the middle of the sun looks brighter than the edges, because the outer rays come to us obliquely, through a greater length of sun's atmosphere than those which come straight through it from the middle of the sun; and that is not the case with the moon which has no atmosphere. Bright patches or clouds, called *faculæ*, are also seen on the left side of spots, as if they were left behind : the spots are called *maculæ*. The earth's atmosphere decreases so rapidly in density, that half its mass is within $3\frac{1}{2}$ miles above the sea ; and at 80 miles high there can be practically little atmosphere.† Cold increases upward, as you get further from the heated earth, at the rate of 1° for 400 feet.

The only thing like a real discovery as to the composition of the sun is that the photosphere contains some of the same materials as the earth, in an incandescent state. For when a ray of sunlight is spread

---

* Appendix, Note V.   † Appendix, Note VI.

out into a *spectrum*, or band of rainbow colors, by
sending it through a *prism*, there are to be seen in it
a multitude of dark lines of various thickness and at
irregular distances, but always the same in every solar
spectrum through the same kind of prism; they are
called Fraunhofer's lines, because he first completely
investigated them, though they had been observed
before by Wollaston. Again, when the light of any
inflamed metallic vapor is sent through a prism, its
spectrum consists of one or more bright colored bands
or lines, sometimes as many as sixty or seventy, as in
the spectrum of iron. But if you send the white light
of an electric flame or any white hot but not vaporised
metal through that same metallic vapor, those bright
lines are immediately turned into some of the dark
ones of the solar spectrum, which then lights up the
spaces left dark in the vapor spectrum with the usual
rainbow colors. And if the vapors of any number of
metals are burnt together, each will keep its own set of
lines, bright so long as there is no purer or more complete light behind it, and comparatively dark as soon
as there is. Hence it is almost certain that Fraunhofer's lines in the solar spectrum are caused by the
vapors of some of our metals existing in either the
atmosphere or the photosphere of the sun.

Iron and some of the secondary metals are thus found
in the sun, but none of the other common or principal
metals. In like manner some of them appear in some
of the stars and others in other stars; which may account for their different tinges of color. For every
color is due to the absence or stoppage of some of the

rays which go to make up a perfect, or pure, or white
light. Things look red which stop or absorb all or
most of the rays which fall on them, except the red
ones. Therefore the question which is sometimes asked,
whether colors exist in the dark, is nonsense ; and you
might as well ask, does light exist in the dark? Therefore also it is that white objects are more visible in a
dim light than dark or colored ones ; for a white surface is one which reflects a good many, at least, of all
the rays which fall upon it, but a dark one reflects very
few, and a colored surface reflects only one kind of rays,
absorbing all the rest ; as you may easily see by feeling
them under a hot sun, when a white board remains cool,
while a dark one is very hot, from the rays which it
has absorbed.

Mr. Nasmyth, who has retired from engineering to
astronomy, with the fame of having added to the powers of man by inventing the steam hammer, and other
persons too, have observed that the sun is covered with
bright clouds of the form of willow leaves, which over-
lie each other in all directions and sometimes close up
over a spot. These bodies are also called ' rice-grains'
and ' pores,' according to the fancy of different observers, and they are now admitted to be of more irregular
forms than was represented in some of the early pictures of them. Sir J. Herschel thinks that they are
not clouds in the sense of being any kind of vapor,
but collections of loose solid matter, for which I do not
know any better word than ' fluffy ;' and that they float
in some highly heated gas or atmosphere which makes
them so hot as to be luminous, though it is itself invisible.

HEAT OF THE SUN.

You will find in Herschel's Astronomy an account of the curious speculations which have been made as to the generation of heat in the sun, and also some calculations of the amount of heat which it sends out in any given time. One is, that each square yard of surface gives out every hour rather more than would be got by burning six tons of coal; and from that you may deduce the following calculation. Six tons of coal weigh about as much as seven cubic yards of the sun's average substance: that is, if the sun were a great coal he would have to burn down seven yards an hour to give out the heat that he does; or 35 miles deep would be burnt off him in a year, which is about a 4000th part of his whole bulk. The consequence is that the sun would have been burnt away into ashes, which would burn no more (though all his weight would remain there), in 4000 years. We may be quite sure then that the sun's heat is not kept up by the burning of his own substance, at any rate.

Another mode of keeping up the sun's heat has lately been imagined, which is remarkable for its ingenuity if not for its probability. It is that the heat is due to a constant shower of meteoric stones falling all round the sun with the velocity they would fall with, if they once got near enough for his attraction to overpower their velocity in another direction. You may soon satisfy yourself that blows will produce heat by holding a nail in your fingers while you hammer it well on an anvil: in fact it can be hammered hot

enough to light a match. It is calculated that if stones
as heavy as granite fell all over the sun 12 feet thick
in a year with the greatest possible velocity (384 miles
a second) from the sun's attraction, they would maintain the actual heat; or if they fell with the more
probable velocity of small planets dragged out of their
orbits sideways into the sun, or 270 miles a second, 24
feet of thickness a year would be required. And as
granite is about 2½ times as heavy as the average matter of the sun, that would be equal to an addition of a
mile to the sun's diameter every 50 years, or an increase of the sun's weight or mass by about a 7000th
part in the 2000 years since tolerably accurate astronomical observations began to be recorded. And that
is enough to have shortened the year by more than
half an hour, according to the law by which the length
of the year depends on the sun's mass and distance, as
you will see hereafter. But the year has not shortened; for even the four seconds mentioned at p. 57
are no shortening of the absolute revolution of the
earth, but are only due to the tilting of the ecliptic.

This last objection certainly would fail, so far as the
earth is concerned, but would still hold as to Mercury
and Venus, if all the meteors or little planets which
fall into the sun are equally scattered round the sun in
a sphere within the orbit of the earth; because then
the attraction of the meteors and sun together, on
bodies outside them all, is the same as if they were
concentrated in the sun (p. 30). Accordingly it is next
supposed that this store of meteors is a certain nebulous and faintly luminous mass called the *zodiacal light,*

sometimes looking like the Milky Way, which surrounds the sun in the form of a lens, or very flat spheroid (but not a sphere) extending nearly or quite as far
as the orbit of the earth. Then it is assumed that a
sufficient number of these asteroids are continually
missing their way (as we may say) and getting dragged
into the sun and swallowed up for food to maintain
his heat by their concussion. The cause of their so
missing their way is imagined to be a certain resisting
medium called *æther*, which we shall see afterward
that there really is reason to believe fills all space, and
must in time gradually contract the orbit of everything
which moves in it.

Not that there is the least evidence of any solid body
in the universe having yet been at all affected in its
orbit by such resisting medium; though one comet is
believed to have had its period shortened in that way.
You will see afterward how extremely unlike comets
are to granite or meteoric stones in density. Nor do
the comets themselves, though very sensitive to disturbance by reason of their lightness, appear to be
ever disturbed by passing right through that mass of
zodiacal light, about 180,000,000 miles wide, full of
real or supposed meteors or asteroids. The crossing of
two streams of meteors by the earth every August and
November gives no help to the theory of the sun being
supplied from a store within the earth's orbit; especially as they are now supposed to have orbits reaching
beyond Uranus, and not at all likely to fall into the
sun (p. 221). Every now and then there have been
flashes of light in the sun, which may possibly have

come from some such concussion. Otherwise there are no visible indications of this rain of stones upon the sun; but perhaps that could hardly be expected; for if a whole year's supply came in one piece, it would be smaller than the moon, which could not be seen at that distance in the brightness of the sun. In this state of things it is not surprising that the meteoric hypothesis is not yet accepted by astronomers as the real explanation of the sun's heat, though no more probable one has been invented.

Leaving these speculations into unknown causes, Sir J. Herschel gives us another measure which has been experimentally obtained of the quantity of the sun's light and heat by himself and others; viz., that the light of any piece of his surface is 146 times greater than that of an equal surface of lime under the oxy-hydrogen flame, which gives the most intense light and heat we can make; and that the heat is enough to melt fourteen yards thick of ice, laid all over the surface of the sun, in a minute. From this you may easily calculate by the rules at p. 24, that that shell of ice round the sun would be about three-fifths of the size of the earth; and therefore the *whole* heat of the sun is enough to melt an earth of ice in less than two minutes. And it would boil all that water in two minutes more, and turn it all into steam in a quarter of an hour from the time it was first applied to the ice. Of course the earth receives only a very small part of all the heat sent off by the sun; indeed less than the 2,000 millionth part, and all the planets together only a 227 millionth. So although we have so much diffi-

4

culty in finding out how the sun's heat is kept up, he
can afford to waste (as far as we know) many million
times as much as all his satellites can use.   The earth
receives as much heat in a year as would melt 100 feet
thick of ice, or boil 66 miles deep of water, all over the
globe.

Latent and Specific Heat.—These figures require a little
explanation.   It takes nearly as much heat to turn ice
into ice-cold water as to raise that water afterward to
212° or boiling : or a pound of boiling water will melt
very little more than a pound of ice.   The exact heat
required to melt the ice is 143°.   Therefore water is
said to contain 143° of *latent heat*, which is merely
employed in keeping it fluid, and does not disclose
itself by the thermometer.   And again boiling water
will swallow up 980° more heat before it is all turned
into steam, and so that is the latent heat of steam.
Consequently it takes nearly half as much heat to melt
a piece of ice as to boil it, and just nine times as much
to boil it all away into steam as to melt it, or five and
a half as much as to make ice-cold water boil.   But
for this remarkable law of nature rivers and ponds and
cisterns would often be frozen to the bottom in a single
night, and remain so as long as the air is a single degree below freezing; and as soon as it rose the least
above 32° all the ice and snow that had fallen would
melt suddenly, and produce destructive floods; and
steam would be unmanageable and almost useless.
Whereas, if you melt a pot of lead it falls into fluid
nearly all at once, not like a lump of ice put into warm
water; and in like manner lead, or cast iron, or mercu-

ry (at 40° below zero), get solid nearly all at once, because they have so much less latent heat than water:
that of lead is only 9°.*

Two other causes prevent water, especially water
only open at the top, from freezing too quickly. One
is that it takes much more heat to raise a pound of
water 1° of a thermometer than a pound of anything
else: thirty times as much as mercury or lead, solid or
fluid; twice as much as oil, and half as much again as
spirits of wine. Therefore they call *the specific heat*
of water thirty times that of mercury, and so on. And
again, the steam of water expands far more than any
other steam. The other reason is that water (like antimony,† iron, and a few other things which crystallize
in freezing) expands instead of contracting before it
freezes, as well as when it is heated, and so the ice and
very cold water keep at the top. Besides that, it *con-
ducts* heat very slowly. The hot water in boiling rises
bodily, and in that way heat is soon *conveyed* through
the mass. Therefore the sea gets heated slowly and
gives out a great deal of heat in cooling, and so equal-
izes temperature.

You may have read in the life of George Stephenson
how he astonished another scientific man by telling him
that his railway engines were driven by the sun's heat
' bottled up ' in the earth from perhaps millions of years
ago. Probably he did not know that the same idea
had been published before, in one of the early editions

---

* See Chambers' Chemistry. In some books latent heat is confounded with
specific.

† Types could not be cast sharp enough to use till that was discovered.

of Herschel's Astronomy. But it is a fact which should be borne in mind while we are reflecting on the business of the sun in the order of nature, that every plant and tree and bit of grass that grows is raised by the sun's heat. And as coal is known to be the condensed vegetables of former ages, chiefly of gigantic kinds of ferns, which could only grow in a very hot and damp air, it is certainly true that every bit of coal or wood which we burn now owes its condition and power of giving out heat again to the heat which it borrowed from the sun, whether twenty years ago or twenty millions.

Moreover you should understand that every force in the world,—that is, every power to lift a weight, from your own foot off the ground to every ton lifted by a steam engine, and all the millions of tons of leaves and wood added to the trees every year, and the clouds full of rain lifted from the sea and rivers by evaporation, as much as the steam off a boiling pot, is all due to heat ; that is, ultimately to the sun. In short, heat and force are now well known to be convertible into each other. The only force which may seem an exception to that rule is the force which lifts the tide or makes the great tidal wave keep rolling round the earth after the moon and sun, as I shall explain afterward ; but that also comes from the sun's and moon's attraction, and the original impulse with which the earth and moon were first set going, both rotating and revolving in their orbits ; for the planets could no more set themselves originally in motion than they could create themselves out of nothing. Those who wish to pursue this subject further should read Mr. Grove's 'Correlation of Forces,' which will open to them new views of the ultimate identity of all the forces in the universe; but still leaving the great force of all, the attraction of all matter for all other matter, standing out as an independent law of nature, imposed at the creation to keep the universe in order.

The sun's rays heat the air very little, and dry air not at all; and the rays which have come through the slightly damp atmosphere heat glass also very little, though glass stops a good deal of the heat from a fire, and damp air keeps heat from radiating away. But for the moisture of the atmosphere we should be exposed every night to the cold of space, 490° below zero, except so far as the heat retained in the earth helped us; and that would rapidly radiate away. A mirror which reflected all the rays it receives would not be warmed at all. When the sun's rays fall on ice and snow they are chiefly employed in melting it, and so the heat becomes latent and is not given out into the air, which is chiefly warmed by contact with the earth. The sun's rays part with about one-third of their heat in passing through the atmosphere, and probably one-sixth of it in the last 4 miles.

NATURE AND VELOCITY OF LIGHT.

Although we do not know how light and heat are generated in the sun, we know that they are to a great extent identical and come together; and we know how long they take to come; and that they are not any substance, or fluid, or gas, emitted from the sun, like the atoms or vapors of scent which are thrown off from odoriferous bodies without diminishing their weight sensibly; but vibrations, or undulations, or little waves, generated somehow by the sun or a fire, in what is called, for want of a better name, the *luminiferous æther,* which is now believed with good reason to pervade all space, and perhaps even the interstices between the atoms of which solids and fluids are composed. This grand discovery of *the undulatory theory of light* was published by the late Dr. Thomas Young of Cambridge in 1802; and after being denounced as paltry, absurd, contradictory, mathematically impossible, and many other things, by the Edinburgh Review,* is now universally received as the only theory that explains all the known phenomena of light, and is said by Sir J. Herschel to ' suffice, if it stood alone, to place its author in the highest rank of scientific immortality.'

It is fair to mention, however, that this theory had been suggested, but not proved, long before by Huyghens, and rejected by Newton for the emission theory. I should remind you also, that the undulations of æther which make light and heat have this material difference from the undulations of common air which make sound, that they can only travel in straight lines; except when they are diverted by passing from one transparent medium into another of

---

* And that twice over, in articles in the 1st and 5th vols. of that Review, well known to be Lord Brougham's. Dr. Young also completed the theory of the tides, which had been left imperfect by Newton, Laplace, and other great mathematicians, and discovered the mode of interpreting Egyptian hyeroglyphics: he was also a considerable scholar. He died in 1829, fifty-six years old, two years after Laplace, who was seventy-eight. See his Life and Writings, edited by Dean Peacock.

different density, or when they are reflected by a mirror. The reason at the bottom of this is that light vibrates *across* the line in which it travels, like the waves you send along a rope by shaking it at one end; but those of sound go backward and forward, as the waves run along a field of corn under the wind, and also spread all round like those from throwing a stone into a pond. But light vibrates in all directions across every line in which it travels, and not only in one plane like the vibrations of the rope; and those lines radiate in every direction from each luminous point, so that the faces of the waves advance like a sphere continually expanding. But it is impossible to explain here how those results follow.*

Light is itself invisible except by its effect in illuminating things on which it falls. If you look into one end of a blackened tube with the other end closed, and a slit or hole in the side for light to come through, you see no stream of light, but only a little faint reflection from the black surface. But if you stick a white ball or a bright wire in, so that the light falls on it, it will shine as usual. What is called a sunbeam shining through a hole in the shutter in a dark room is not light which you see, but the motes of dust shining in the light. The full moon stands in a flood of sunlight, which is dark all round her and invisible, except where it illuminates her surface.

Here let us stop a moment to reflect on the importance of this fact, or law of nature, that light will

---

* You may read with advantage the explanation of light in Sir J. Herschel's 'Familiar Lectures,' lately published.

only travel in straight lines (except when it is artificially diverted), while sound can turn round corners, only rather weakened by it. Not merely the whole science of measuring, but the power of judging of sizes, distances, and positions of everything we cannot touch, depends on the lines of sight going straight. If the case had been reversed, we should have lost a great deal by every sound being stopped the moment anything came between it and us, and gained nothing; while if we did not see straight the whole world would be in confusion, and it is hardly too much to say that the human race could not exist.

There is also a vast difference between light and sound in their rate of travelling. The waves of sound go only 377 yards in a second, while the earth itself goes 18·2 miles, and light ten thousand times faster than that; while electricity (which again is probably another kind of vibration, of the solid atoms of bodies, and certainly not a fluid) runs along a wire about half as fast again as light.* So if the earth were a cannon ball, shot at the sun from its present distance, with the velocity it now travels with, and the moment of explosion telegraphed to the sun, they would get the telegram there in about 5 minutes, and see the earth coming in 8 minutes, and would have nearly two months to prepare for the blow, which they would receive about 15 years before they heard the original explosion. This is merely taking the sun as a target to be shot at, without regard to his power of attracting the earth at the final rate of 384 miles a second.

---

* Appendix, Note VII.

You see then that 'rays of light,' or a 'beam of light,' which we still speak of for convenience, are not, as they used to be thought, emanations of anything in mathematical straight lines of 'length without breadth;' but are merely the straight course in which vibrations run along the æther in every direction ; and a luminous body is one which somehow or other has the power to excite those vibrations of light—and heat, which (as far as we know) always attends it.

The general connection of light and heat is obvious enough : it required a great philosopher to separate them as they come together from the sun. It is not merely that things can be moderately hot without being luminous (though not very hot), or that some lights have little heat ; but Sir William Herschel discovered that every ray of sunlight can be split up into visible and invisible rays ; the visible ones consisting of the seven colors of the rainbow, as Newton had discovered long before, and the invisible rays consisting of heat, even hotter than the red or hottest end of the visible spectrum or band of colors, which appear when light passes through a prism or wedge-shaped piece of glass. Besides these there are other invisible rays beyond the violet or cool end of the spectrum, the effects of which can be made visible by chemical contrivances, as in photography ; as the invisible hot rays can be concentrated by a lens or burning glass, so as to set things on fire.

And they all come out in this order, beginning with the most refracted : chemical rays (invisible), violet, indigo, blue, green, yellow, orange, red, and then hot rays invisible. Those invisible hot rays would be hotter still, but that they are the rays which are chiefly stopped by our atmosphere and go to warm it. From an electrical light the invisible rays are much hotter than the visible ones, not having many miles of air to pass through.* We shall see more about this refraction of light into colors in the chapter on telescopes.

Resisting Medium.—But now comes a surprising astronomical result of this theory of light consisting of undulations of some æther or medium filling all space, but far thinner than the most rarefied air we can produce, so thin that we can neither weigh it nor test its presence by any positive evidence as yet. Still, if it exists it must have some density; and then space is not absolutely empty, and the earth and planets do not move through space absolutely without friction and resistance, any more than a stone or a pendulum moves through the air without friction, only that the resistance of the æther is immeasurably small. The only one of the heavenly bodies which is even probably suspected to have been sensibly obstructed by this resisting medium, is a small comet called Encke's, which returns every $3\frac{1}{8}$ years, and whose period shortens about $2\frac{1}{2}$ hours in every revolution : and therefore its distance from the sun is shortening too. Sir J. Herschel however suggests another possible cause for this, in the comets leaving part of their tails behind them every time they pass the sun, as some of them manifestly do. The tail matter is repelled, not attracted by the sun, and so there is more attraction left on the remaining

* See Tyndall on Heat, § 308, etc.

matter of the comet, which makes it move in a smaller orbit, and therefore faster. The matter which afterward forms the tail appears not to be emitted on the far side, and therefore cannot produce a recoil toward the sun, or a diminution of the orbit. Encke's comet however has no tail. The general opinion of astronomers is in favor of the resisting medium, though more from the phenomena of light than from the acceleration of a single comet.

Acceleration by obstruction will seem strange; but you must remember that a body may perform a smaller orbit in less time though it goes fewer miles per hour; and the only velocity visible among the heavenly bodies is their velocity of revolution. The resistance does diminish the actual or *linear* speed, and thereby diminishes the centrifugal force which resists the sun's attraction, and so the orbit is contracted and the time of revolution quickened. And though comets are so thin and light that we can see through their bodies, and they can pass among the moons of Jupiter without sensibly disturbing them, still they are composed of matter and obey the same laws as the planets; and if the friction of the luminiferous æther can drive one comet a mile nearer to the sun in any time whatever, it is absolutely certain that in some time or other, though it may be an inconceivable number of years off, the earth itself will fall upon the sun—and be again melted by the blow: as its original state of fusion is conjectured to have come from the sudden bringing together of its atoms.*

---

* Appendix, Note VIII.

We need not be alarmed at this conclusion. All the heavenly bodies had a beginning, for they were created, and they may as naturally have an end. The earth, we know, will have an end, 'and the elements shall melt with fervent heat.' By whatever means the sun's heat is maintained, no philosopher doubts that it is by the using up of something, or the destruction of some motion, for heat and force are universally convertible.* And so, although the law of the stability of the universe is true, and all the bodies in the solar system can recover from the temporary disturbances which are always taking place by their mutual attractions, it is almost equally certain that in the lapse of ages permanent changes take place also. Some conspicuous stars have gone out since men began to observe them, and the stars are not little things like the earth, but many of them probably as big as the sun, and some certainly much bigger, as we shall see hereafter.

The ultimate destruction of the solar system, however remote it may be, is a tremendous consequence to follow from such an apparently minute fact as that light consists of very small vibrations of an immeasurably thin æther. But thus are the laws of nature linked together. Perhaps there may be some counteracting and preserving force in operation not yet discovered. And now leaving these contemplations of the future, and returning to the actual velocity of light, we shall find that its latest measure confirms the late discovery that all the dimensions in the solar system had been rather overrated by astronomers.

---

* Appendix, Note IX.

NEW DIMENSIONS OF THE SOLAR SYSTEM.

From 1769 to 1862 the sun's mean distance was considered to be 95½ million miles, and so you will find it still in most books of astronomy. Mr. Airy, the Astronomer Royal, said in his lectures delivered in 1856, 'there is no probability of an error of half a million of miles' in this distance. Yet within six years it became almost certain that there was an error of four millions. The error was even then suspected by Professor Hansen, of Gotha, from the 'parallactic inequality' of the moon appearing to be wrong; and the suspicion was confirmed by certain observations of Mars in 1862, which will be explained at p. 209.

For you must understand that the measuring of the actual distances in miles is difficult, though measuring the relative distances is not. It is easy enough for astronomers to make out, for instance, that the distances of Mercury, Venus, and the earth from the sun are nearly in the proportions of 4, 7, and 10; but the difficulty is to find a scale to measure any one of the distances by. A convenient scale is the diameter of the earth itself; but that is so exceedingly small compared with the smallest of those distances from the sun and other planets, that it is very difficult to apply it accurately. I will explain hereafter how it is done; just now we are only inquiring into the velocity of light. About the same time that astronomers began to suspect that the dimensions of the solar system had been overrated, some new experiments on the velocity of light were made in France with machines on the principle of one which had been invented by M. Foucault or Mr. Wheatstone for electricity ; and it was found that the velocity formerly accepted, of 192,000 miles in a second, had been overrated to about the same extent. Then came further astronomical observations, which have all tended to confirm those of 1862 ; and it seems now to be tolerably clear that the velocity of light cannot be more than 184,000 miles a second, and the sun's distance about 91,404,000 miles,* or just one 24th less than the old amount.

On the 8th of December, 1874, and again in 1882, will occur an event of great astronomical importance, which has not happened since 1761 and 1769, when some mistake was evidently made, and will not happen again for more than another century, that is, a transit of Venus as a small black spot over the face of the sun. For that phenomenon is considered to afford more accurate means than any other of measuring the sun's distance compared with the diameter of the earth : not that a single observation of that kind can prevail over the concurrence of evidence that now exists in favor of the reduced distance; but any confirmation thereby will be decisive.

With the distance of the sun, every distance and diameter in the solar system (except of the moon, which is measured independently, being tolerably near), and even the weights or masses of the sun and planets have to be altered. First of all, their real diameters depend at once upon their distance : for all that we can say of them is that they appear a certain width ;

---

* Appendix, Note III.

that is, the two lines of sight from their opposite sides to our eye enclose a certain angle; but a shilling near your eye looks as big as a crown further off; and so we have to wait till we know whereabouts the sun has to be placed between the two lines of sight which measure his apparent diameter before we can say what real diameter it corresponds to. Therefore it is plain that the diameters have to be reduced, with the distances, by a 24th of the old amount; and as the bulk varies with the cube of the diameter (as I explained at p. 24) the bulks have to be reduced in the proportion of 24^3 to 23^3; which you will see, if you calculate it, makes a reduction of ·12 or nearly one 8th in the old bulk of the sun and all the planets except the earth.

The reason for altering their masses or weights is not so evident; for you must remember that we know nothing of their density except through their size and weight; and so we cannot say that the weight is to be reduced an 8th because the diameter is reduced a 24th. Still it has to be done for quite another reason, and as it happens, in just the same proportion. One of the things proved by Newton was that the mass of the sun must bear a fixed proportion to the cube of its distance from the earth and the square of the time of the earth's revolution, as you will see at p. 250. Strictly speaking, it is the mass of the earth and sun together, but the earth is so small in comparison that it may practically be left out of the calculation. And as there can be no considerable mistake about the length of the year, we may say that the mass of the sun bears a fixed proportion to the cube of its mean distance. Therefore, when we find the distance has been overrated by a 23rd, we have to reduce the mass of the sun in the proportion of 24^3 to 23^3, or 100 to 88, or nearly 8 to 7.

Here I stop for a moment to remark how much better the old, and the mathematical, habit of using vulgar fractions in small figures is than the modern vulgar habit of *per-centing* everything. It is very easy to remember, when you have once learnt it (which few people seem to do), that a reduction of a 24th corresponds to an increase of a 23rd, a reduction of one-third to an increase of one-half, and so on; or that any two numbers, such as 2 and 3, which differ by one-half of the smaller number, differ by one-third of the larger. But no such rule can be expressed in decimals. People nowadays, especially commercial ones, will not even talk of a half or two-thirds of anything, but of '50 per cent.,' and '66 and two-thirds per cent.;' and sometimes they talk of a thing costing 100 per cent. *less*, as if it was the converse of costing 100 per cent. *more*, instead of being nonsense; for they do not mean that it costs nothing, but half as much. If they would condescend to use vulgar fractions whenever they can be expressed in small figures, and especially when the numerator is 1, they would save a great many words and make fewer mistakes. In some cases decimals are best.

Then comes the question, why the masses of the planets are to be reduced for their new distances. Jupiter and the others which have moons affect their moons or satellites as to time and distance just as the sun affects the earth; and therefore, as the satellites, distances have to be reduced with the rest, the masses of their attractors or *primaries* must be reduced like the sun's. As to Mercury, Venus, and Mars, which have no moons, their masses were calculated from their powers of disturbing, that is, attracting their neighbors and comets, and therefore their masses also have to be diminished as the cubes of their distances.

But how is that reduction to be measured or expressed? By the only unit or standard of weight that we have for all the solar system, that is, the earth. It is much easier, as I shall show you afterward, to weigh the sun in *earths*, than you have already seen it is to weigh the earth in tons, or to compare it with an equal bulk of lead or water. Therefore if we keep the old figures for the proportions between the masses of the sun and planets, while we reduce the sun from the old figure of 359,500 to 316,560 times the earth, we shall have reduced them all. Still you may ask one more question: Why is not the proportion of the sun to Jupiter to be reduced as the proportion of the sun to the earth is? Because the mass of the sun was not obtained from Jupiter, nor that of Jupiter from the sun. And that would be not reducing Jupiter at all; for he would be first reduced one 8th with the sun, and then increased the same one 7th again, to reduce the sun's proportion to his. Neither the sun nor Jupiter is the unit of mass for the solar system, but the earth is.

Attraction is Instantaneous.—As I have given the times which electricity and light would and do take to

reach the earth from the sun, I ought to add that the
force of gravitation alone seems to be instantaneous.*
All the calculations of the motions of the most distant
bodies, no less than of the nearest, assume that it is,
and the results justify the assumption. See the note
on this at the end of Sir J. Herschel's lecture on the
sun.

---

* Appendix, Note X.

CHAPTER III.

If you ask half a dozen people how big the moon looks, you will probably get as many different answers, varying perhaps from a shilling to a large round table. The truth is that the question has no meaning, because a shilling near the eye looks as large, or makes the same angle with the eye, as a plate further off, or as the moon further still. But then some people say the moon looks bigger than the sun, others smaller, and others about the same size; and that has a meaning, and can be wrong or right ; for we can compare the angle which each of them makes with the eye, and say which has the *real appearance* (quite a different thing from the reality) of being the largest. Without any measuring at all, there happens to be a ready way of knowing that they really appear about the same size, sometimes one a little larger, and sometimes the other. For the moon quite covers the sun in a *total* eclipse, but leaves a bright ring all round it in an *annular* eclipse. In the former she appears at least as large as the sun, being at her nearest to the earth : in the latter a little less, being further off.

But as soon as we take the distance of the sun and moon into account, we begin to see what a little thing the moon is. Her distance from us varies from 225,-720 to 251,940 miles ; for she moves in a more elliptical orbit round the earth than the earth does round the sun, the eccentricity being one 18th, or above three times the earth's. The mean distance is 238,830 miles, or 30·1367 times the earth's diameter. The moon's diameter is only 2165 miles; therefore the earth is 49 times as big, and the sun more than sixty million times as big as the moon. If you take a ball of an inch diameter for the earth, the moon will be little more than a quarter of an inch, or about the size of a pea 2½ feet off, and the sun a globe of 9 feet diameter standing 320 yards off. Moreover the density of the moon is only ·6 of the earth's, or 2·4 of the sun's density; for the earth is 81·5 times, and the sun nearly 26 million times as heavy as the moon; and gravity there is only one-sixth of what it is here, although her surface is only about one-fourth as far from her centre as ours is.

It is curious that the sun's distance is nearly as much greater than the moon's (383 times) as his diameter is (390 times); also that the moon's distance is 110 times her diameter, and the sun's 108 times his diameter, which again is 107 times the earth's diameter.

The moon is so near that we can see the nature of its surface to be extremely different from the earth's surface. It is covered in many places with craters of old burnt-out volcanoes, far larger than any on the earth, and with the heaps of ashes which were thrown out forming mountainous rings around them. Indeed it seems that all volcanic action is not extinct, as some persons have supposed that one crater called Linné, 7 miles wide, has been lately raised from a hollow into a

hill.* There are also wide plains called seas: but not
of water, for there are no signs of any; nor of any atmosphere affecting the light near the edges of the
moon,† as there are in the sun and planets. The reg-
ularity and distance of the rings from the craters, and
the height of the mountains, would depend on the volcanic force and the inclination at which the matter was
thrown out, as there was no wind to stop or spread the
ashes; and that force would be practically greater on
account of the weakness of gravity and the quicker
cooling and contraction, from the moon's smallness and
the want of any atmosphere to check radiation away
of the heat into space.

A telescope which magnifies 1000 times shows the
moon as we should see it 233 miles off without a telescope; for we have to measure between the two surfaces, not the centres. A town a mile wide, at that
distance, would look nearly half as wide as the moon
does, and a building of the size of York Minster would
look larger than Venus.‡ We see the shadows of the
mountains clearly when they are turned sideways to
the sun.

Rotation of the Moon.—It is one of the peculiarities of
the moon, and (as far as we can see) of the moons belonging to other planets also, to keep the same face always toward us, as if it were set fast upon a stick
reaching to the earth. This is accurately described by
saying that the moon turns on her axis exactly in the
same time as she revolves round the earth. You may
very likely think that the moon's keeping the same face

* Appendix, Note XI.   † Appendix, Note XII.   ‡ Appendix, Note XIII

to us proves that she does not turn on her axis at all; and every now and then somebody undertakes to prove that all the astronomers are wrong in saying that the moon rotates or turns upon her axis; as others undertake to prove that all the mathematicians in the world are wrong in saying that the diameter of a circle bears no exact proportion to its circumference or area.*　I know it is as hopeless to convince them as it was the man who insisted that his head had been twisted round. But for the information of those who wish to learn I will give a few of the innumerable proofs that can be given of the moon's rotation : see also p. 57.

First, then, the deniers of it forget, or do not know, that if they were on the sun, or anywhere else in space except just here inside the moon's orbit, they would see opposite sides of her every fortnight; and the proposition that she does not turn would then appear as absurd to them as the contrary does now, and with much better reason.　Perhaps the mistake arises from people thinking of the moon's axis as a wire stuck through it, and forgetting that if it were, the question would remain whether the wire is turning on its own mathematical axis or not.　If you put a wire upright through a ball with one side painted white, and bend it into a horizontal tail which you can hold always pointing to one side of the room, while you carry the ball round you with the white side always toward you,

---

*I mean no proportion measurable by any finite number of figures, or by any geometrical construction with rule and compasses.　Mathematicians know very well that it can be done mechanically by a cycloid, or the curve traced by any point in the circumference of a wheel rolling on a straight rail; for the base of the cycloid evidently — the circumference of the wheel.

it will plainly enough turn on the axis, because the axis has been kept from turning. But if you keep the tail of the wire always toward you, the ball does not turn on that axis because it has turned itself just as much.

Or you may lay a book on a revolving table, at some distance from the centre. The deniers of the moon's rotation say the book does not turn on its own axis because it keeps the same face to the centre of the table when you turn it round. But that can have nothing to do with its distance from the centre. Keep gently moving it nearer to the centre while the table revolves, and at last it will be visibly turning on its own axis and nothing else. Then at what moment did it suddenly start from no rotation at all into a rotation at the same rate as the table?

Once more, set a cotton reel with a hole bored across it, upright and fast on a nail at the end of a stick which can revolve round a pin stuck in a table, as the moon revolves round the earth; and put a string in the hole as if you were going to wind it round the reel. Then turn the stick round, holding the string in your other hand: it will wind once round the reel for every revolution of the stick, just as much as if it had stood still and you had turned it round between your fingers.

So much for the fact of the moon's rotation. But you may prove also that a force of rotation must have been originally impressed upon it to make it rotate even at that slow rate of once a month. Hang a globe or anything else, so marked as to distinguish the sides, by a thin string from the end of a stick held in your hand, till it has got quite steady: then turn round, carrying the stick with you so as not to shake the globe. You will find that by the time you have gone half round, the side which was facing you at first has got furthest off, but still faces the same side of the room. Why? Because there has been no force impressed on it to make it turn, and nothing ever moves except in obedience to some force.

Some astronomers believe that the reason why the moon keeps the same side toward the earth is that her far side is heavier, or 'her centre of gravity 33 miles beyond her centre of figure;' which it is said would keep the heavy side furthest from the earth by centrifugal force, as the heavy end of a stick will keep furthest off if you swing it round by a string tied to its middle, not to its centre of gravity.* But that analogy seems to me to fail; for there is nothing to make the moon turn upon her centre of figure: on the contrary, a body that is free to turn will always choose an axis through its centre of gravity to turn upon; and if you swing the stick by its c. g. the heavy end will not keep furthest off any more than the light one. It is remarked truly that the c. g. being nearest to the invisible side of the moon would make all the water go there, if there is any, and so account for its non-appearance: but there seems to be no reason for believing that there is any.†

Again it is said in other books that when the moon

---

* Herschel's Outlines of Astronomy, p. 237 (ed. 7 and 8).
† Appendix, Note XIV.

was fluid the difference of the earth's attraction on the near and far sides would make her egg-shaped, as the moon's attraction raises the water of the tides slightly into that shape. And so it would—but not to the extent of 30 miles, as I shall show you at p. 134. If she really has that shape, some better cause for it has to be discovered yet, which must apply to the moons of Jupiter and Saturn also. Mr. Adams does not consider it proved by any of the calculations which are supposed to prove it. But it may still be open to othe. proofs, and I shall have to mention one at p. 104, if the observations there stated are correct; and if so, the c. g. being nearer the far side will account for that side keeping furthest off, not by centrifugal force but for another reason.

Another illustration will be useful for another purpose. If a man is waltzing with a little girl, each turns round the other, and on their own axes too, facing all the four walls of the room in succession; but besides that, the man being the heaviest makes a smaller circle round the girl than she does round him: that is, he swings her round him. And so the earth goes round the moon as truly as the moon round the earth, only in a circle as much smaller as the earth is heavier : that is, 81·5 times. The point between them which remains fixed, if we forget their joint motion round the sun, is their centre of gravity, which is 81·5 times nearer the centre of the earth than of the moon; and as the distance of the moon is only sixty times the earth's radius of 3963 miles, the centre of gravity of the earth and moon is 2918 miles from the earth's centre.　So if you fix two balls on the ends of a stick five feet long, one four times as heavy as the other, and whirl it into the air, they will turn round the centre of gravity, four feet from the light ball and one foot from the heavy one.

The cause of the moon keeping her distance from the earth is simply the centrifugal force of both of them.　If the light ball just now described can slide along the stick, but is connected with the other by a string, and you whirl them into the air, they will keep the string stretched by their centrifugal force.　The same is the reason why the earth and all the planets keep their distance from the sun, subject to the variations due to elliptic motion, as I shall explain hereafter.

Calculations of the moon's motion are simplified by ' reducing the earth to rest,' as it is called ; which might be done by assuming another equal moon at the same distance on the opposite side of the earth, so that the common centre of gravity and of motion would always be the earth's centre.　But this hypothesis would make the mass of the whole system too great, and also increase the time of revolution ; and the more correct way of reducing the earth to rest is to suppose the mass of the moon transferred to the earth, leaving her only an empty shell ; for the attraction on, but not of, a body is independent of its own weight or mass (p. 28).　It is important to attend to this in calculations of the moon's period round the earth, and in weighing the sun thereby, as we shall do hereafter.　But whenever we have to deal with the moon's attraction, as on the equatorial protuberance of the earth in precession and nutation, or on the water of the earth in the tides, we must of course deal with her real mass, though we may still consider the earth at rest, except in rotation, because that alone affects those disturbances.

The Moon's Orbit.—The motion of the moon in space is harder to understand than its motion round what we call for convenience the earth, meaning that point in the earth which is their centre of gravity. You may think it is not so very difficult to see that the moon describes a kind of spiral path with loops in it, round the earth, which goes on in its orbit while the moon turns round it in the same direction from west to east; and the pictures in some books represent it so. But it does nothing of the kind, as I will show you by a thing which you may see any day. If a long pencil were stuck into a railway wheel at any point within the circumference, so as to mark its own path on a wall at the side of the railway, you would see that it is nothing but a wavy line, with the lower bends rather smaller than the upper.* And the moon's path round the sun—and the earth's too in a less degree, is only the regular elliptic orbit of their common centre of gravity with 25 places in it, for full and new moon, a little more and a little less alternately curved toward the sun; for none of the bends are convex, but all only more or less concave, toward the sun.

This is difficult to conceive and impossible to draw

---

* That curve is called a *trochoid ;* but when the tracing point is on the circumference of the wheel it becomes a *cycloid*, which is rather like the long half of an ellipse, but very different in its mathematical properties, which are remarkable in various ways.

on a small scale, but easy to prove by a little calculation of which I will give you the result. The c. g.
of the earth and moon moves through 14° 33' in its
orbit round the sun from one half moon to another; and
such an arc described with a radius of 91½ million miles
bends outward 684,000 miles beyond the chord or
straight line joining its two ends. But the new moon
is only (say) 240,000 miles nearer the sun than the two
half moons, and therefore it is still 444,000 beyond the
chord that joins them: that is to say, the path of the
new moon through space is concave toward the sun to
that extent. It is something like a piece of a circle described with a radius of 150 instead of 91½ million
miles, though it is not really circular. In the same
way you would find that the path of the full moon resembles an arc of about 60 million miles radius, evidently more concave toward the sun.

But in all calculations of the moon's visible place, her
orbit may be considered an ellipse, with the earth standing in the focus, and the sun revolving round them
both. And this elliptic orbit, or its major axis, or line
of *apsides*, or *perigee*, and *apogee*, or least and greatest
distances from the earth, revolve in 8 years and 310·575
days, going forward; whereas the earth's perihelion
takes 110,880 years to revolve, independently of the
precession of the equinoxes, which go the other way.

You may easily see a revolving ellipse if you hang a
weight by a long string to a hook in the ceiling and
send it swinging in any direction, except straight across
like a pendulum or quite round in a circle, which indeed
you would find it difficult to do. If you stand near the

narrow part of the ellipse where the weight passes by
you, and stay there, the ellipse will wheel round and the
weight will hit you before long. The only difference is
that the centre of force in this ellipse is not in the
focus, but the centre, the force being the tendency of
the weight toward the vertical or lowest point ; and the
force is nearly in direct proportion to its distance from
that point, instead of the inverse proportion of the square
of the distance. This ellipse wheels forward, or in the
same direction as the weight, because the force does
not quite vary as the distance ; and the lunar ellipse
similarly wheels forward because the moon's attraction
to the earth is a little diminished by the sun, as will be
explained at p. 152.

The moon's axis, like the earth's, does not stand upright to her orbit, but it is much less inclined than the
earth's, leaning only 6° 39′ from the poles of her orbit,
instead of 23° 28′ ; that is, the moon's equator is so
much inclined to the plane of her orbit. And again
the moon's orbit round the earth is inclined 5° 9′ to the
apparent orbit of the sun round the earth, or the ecliptic.
So we have now got one more plane and one more inclination to consider than we had in dealing with the
earth's equator. It is however simplified by the fact
that the crossing of the moon's orbit and equator always
coincides with the crossing of her orbit and the ecliptic;
though that line, which is called the line of *nodes*, revolves backward 19° 19′ a year, making a sidereal
revolution in 18·6 years, or 6793·391 days, or 223 calendar months, just as the equinoctial points ♈ and ♎,
the nodes of the earth's equator and ecliptic, revolve
.

once round in 25,868 years (p. 52). The node at which the moon goes up to the north of the ecliptic is called the ascending node ☊, and the one where she goes down again to the south is the descending node ☋.

Moreover it is remarkable that the ascending node of her orbit is the descending node of her equator : or at the moment when she rises in her orbit from the ecliptic at the inclination of 5° 9′, her equator is inclined 1° 30′ to it *the other way*. Therefore the ecliptic always lies between the plane of the moon's orbit and the plane of her equator, making the angle 5° 9′ with the former and 1° 30′ with the latter. You will see presently the importance of understanding clearly the inclination of the moon's orbit, and the recession of the nodes : its cause will be explained at p. 156.

Libration.—It is not strictly true that the moon always keeps exactly the same face to us, as I said she did ; for she apparently rolls a little in her orbit, so as sometimes to show a little more on the right side, and sometimes on the left. That is called *libration ;* and it has been ingeniously used to take stereoscopic photographs of the moon, which show the heights of the mountains and the depths of the volcanoes. For a stereoscopic view of anything is got by taking two pictures of it, one seeing a little more of the right side and the other of the left, just as your two eyes do, as you may see by winking them alternately while you look at something not far off ; and that is the way the eyes judge of distance, when it is not too great for any sensible difference in the view which each eye takes.

The reason of the moon appearing so to roll is, that

she does not move in her elliptical orbit with uniform velocity, but does turn on her axis uniformly, and so the two motions do not exactly keep pace with each other. She is sometimes 6° 18′ before, and sometimes 6° 18′ behind the longitude she would have if she moved uniformly : that being her greatest ' equation of the centre,' or difference of true and mean ' anomaly ' or distance from perigee (p. 41). And the libration is the same as if the moon stood still rotating uniformly, and the earth went round her and at the same rate on the average, but sometimes 6° 18′ before and behind its mean place. Consequently that is the amount of the *libration in longitude,* each way; and we see on the whole 12° 36′ beyond the half of the moon's surface ; and indeed nearly 16° by reason of the other inequalities in her motion which will be described hereafter.

But the polar axis also leans 6° 39′ from the perpendicular to her orbit, and sometimes one of her poles leans toward us and sometimes the other ; and so we see also 6° 39′ beyond each pole alternately. This is called the *libration in latitude.* Now the area of a *lune,* or slice of surface of a globe between two meridians, is evidently measured by their difference of longitude; and therefore the surface disclosed by libration in longitude bears the same proportion to the whole moon's surface as 15° 50′ does to 360°, and that by libration in latitude as 13° 18′. Therefore (except that they overlap a little) we may say that we gain by all the four librations together, $\frac{29}{360}$ of the otherwise invisible half of the moon, or we see four 7ths of her whole surface at one time or another.

Shape of the Moon.—Some of the stereoscopes of the moon taken at librations certainly present the appearance of a prolate spheroid pointing toward the earth. This may be an optical illusion or defect; but other observations appear to show that she actually is like an egg with the small end toward the earth and the large end behind. If you look straight endways at an egg, with one eye, you cannot tell that it is not spherical, but you can if you see it a little obliquely. By a careful measuring of angles you might find its exact dimensions, and the more obliquely you can see it the better. Now we can get a view of the moon's small end (if it is one) even more oblique than the libration of 8°, by looking at her from the east side of the earth, or at moonrise, when the libration is toward the west, and in that way gain an angle = 4000 miles divided by 239,000, or about 1° more. It appears (as I understand the description) that the oblique profile of the moon does show her to be egg-shaped; and so much, that unless her density is irregular, her centre of gravity is full 30 miles nearer the back than the front; though you can hardly call the middle of the length of an egg its 'centre of figure' or 'of symmetry.'

And then this consequence would follow: the average mass of the long half of the moon is further from her c. g., round which she must rotate (p. 97), than the mass of the short half; and the earth's attraction on each half, being inversely as the square of its distance from them, will be greatest when the long half points to the earth: which will tend to keep it there; though not strongly enough to prevent the libration of a few degrees, which is due to the moon's force of rotation. But how she got that shape is a question yet unsolved.

Light and Heat there.—Though we see only one side of the moon, the sun sees or shines all round her in the course of her month of 29½ days, from one new moon to another; and one of the consequences of there being no water, and no clouds or vapor or air round the moon, is that the side facing the sun is probably hotter than boiling water, and the dark side for the time is colder than the hardest frost on the earth; in fact as cold as space (p. 42), except so far as it is warmed by the heat remaining from the sunny fortnight. So if there be any living creatures there, they must be able to bear those extreme changes every fortnight, besides doing without air and water.

The moon has no light of her own, and only reflects the sun's. Dr. Wollaston made out that full moonlight is 800,000 times weaker than sunlight, which he found equal to 5563 wax candles at the distance of a foot, while the moon was only the 144th of a candle.* Therefore, as the moon fills about the 240,000th of the visible hemisphere, we should get much less light from a whole sky full of moons than we do from the unclouded sun.

Moonshine gives no sensible heat, even when concentrated by a very large concave mirror into the greatest possible intensity on a thermometer; and it causes cold indirectly; for a full moon is found to drive away clouds, and a light night is always colder than a cloudy

---

* Philosophical Transactions of 1829, p. 20.

5*

one (when other things are equal), because the heat of
the earth radiates away into space instead of being kept
in by the clouds.*

Phases of the Moon.—The moon is full when the
earth is between it and the sun, and therefore we
see all the bright face; except when there happens to
be an eclipse of the moon, of which I shall have more
to say presently. The new moon is when the moon
is between the sun and the earth, so that only the dark
side faces us, and we see no moon at all. The phases
of the moon are nothing but the Greek word for her
faces or appearances. She increases in width, or *waxes*,
from new to first half moon in the form called a *crescent*, with the horns turned to the left, and *wanes* from
second half moon to new as a crescent with the horns
turned to the right. From first half to full moon, and
from full to second half moon, in the second and third
'quarters,' one side is a semi-circle and the other side
is half an ellipse, which the hollow side of the crescent also is. That shape is called *gibbous*. The ellipse
in each case is the oblique view of half of that meridian of the moon which divides the light from the
darkness. If you turn back to the figure of an ellipse
within the circle of which it is the perspective view
(p. 40), the shaded part represents the crescent, and
the unshaded the gibbous phase of the moon.

A day or two before or after new moon you may see
what is sometimes called ' the new moon with the old
one in her arms,' or a bright narrow crescent, with the
rest of the circle just light enough to be seen. That is

---

* Appendix, Note XV.

the reflection of the earthshine back from the moon. When it is further from new, the earthshine appears less and the proper moonshine stronger, and so the earthshine is too weak to be seen.

As the moon has no light of her own, but only reflects the sun's light to us, so we do the same to her. If there were any men in the moon they would see corresponding phases of the earth, which would be full to them when the moon is new to us; and the earth would appear to them nearly thirteen times as large as the moon does to us: it would be sixteen if the diameter of the earth were quite four times that of the moon; for the apparent size of a globe, which is called its *disc*, being a great circle of the globe, varies as the square of the diameter, while the real size or solid content varies as the cube, as I explained before.

Periods of the Moon.—'Period' means the time of performing a journey round. The exact time from one new or full moon to another is 29d. 12h. 44m. 2·87s., or 29·5306 days, or 708·734 hours. But to reach a second new moon, or the line between the earth and sun, the moon has to go rather more than once round the earth, because the earth has not been standing still, but by the time the moon has got once absolutely round the earth we and the moon together have gone forward about 44 million miles in our joint annual journey round the sun. And as the moon goes round the earth in the same direction as the earth round the sun, or what we call from west to east, she has to travel further round the earth to get again into a line between it and the sun. If the earth stood still, the solar period of the moon would be only 27d. 7h. 43m. 11·5s., or 27·32166 days, or 655·72 hours, which is its *sidereal* or absolute period with reference to the stars. The other, of 29½ days, is called the *synodical* period, or simply, a *lunation.*

The synodical period of two bodies revolving round a third, really or apparently, is the time of their all three coming a second time into the same relative position. If the two go the same way, their relative or synodical velocity is evidently the difference of their separate velocities : if they go opposite ways, the sum of them ; and velocities of revolution evidently vary inversely as the periods, or directly as the *reciprocals* of the periods; for 1 divided by anything is called its reciprocal. Therefore the reciprocal of the synodical period is the difference, or the sum, of the reciprocals of the two absolute or sidereal periods. From which it follows by a common sum in fractions that the synodical period of two bodies going the same way, like the sun and moon, is the product of their several periods divided by the difference; or by the sum of them if they go opposite ways, like the moon and her nodes.

A lunation then is the time the moon takes to come again into the same longitude with the sun, called *conjunction*, or to get 360° in advance of the sun, or apparently once round the earth, measuring by the sun. Therefore the moon advances 12° 11′ 27″ more than the sun daily.. And as the earth turns in the same direction, it has to turn that 12° 11½′ more than once round for the moon to cross the same meridian again ; which makes the *lunar day* 24h. 49m. Twelve lunations, or

354·367 days, are called a *lunar year*. Therefore the moon is nearly 11 days older (until the excess is more than 29½) at the beginning of every year on the average. The moon's age at that time is called the *epact* of the year.

Harvest Moon.—But the time of moonrise is by no means uniformly 49 minutes later every day. It varies for the same reason as the time of sunrise, and even more, because the moon goes 5° further from the equator than the sun (see p. 101). The sun rises at the same time for several days at each solstice, but the moon nearly at the same time for several days in every month, from a different cause, in high latitudes like this. To understand this you had better take a globe, and elevate the north pole about 52° above the south horizon. Then if you turn the globe round from right to left, as it really moves, you will see that the ecliptic at one of its nodes or crossings of the equator rises nearly flat, or a good deal of it seems to rise at once, and at the other node it rises at a great angle to the horizon. And you will see that the node which so rises flat is the one with the mark ♈ upon it, or the node of the vernal equinox, which is the ascending node of the sun. If you stick two or three wafers on the ecliptic there, about 12° apart, they will represent the moons of successive nights, which will evidently rise nearly at the same time. And that happens to *some kind of moon* every month, as the moon goes through all the signs of the ecliptic in 27·32 days. But it only happens to one *full* moon in the year, viz., to that one which comes at or nearest to ♈, which is when the sun is at the opposite node ♎, or the autumnal equinox. So the moon rises about full at six o'clock or sunset for several nights together, always within a fortnight of September 23; and that is called the harvest moon.

This uniformity of time of moonrise for a few days in every month is also greatest when the moon's ascending node coincides with ♈, for then the moon's orbit is 28½° inclined to the equator, and still nearer to the colatitude of England, which is about 38°. On the other hand, when the moon's ascending node coincides with the sun's descending node ♎, the inclination of the moon's orbit to the equator is only 18½°, and therefore the 'harvest moon' effect is considerably less.

Moonlight in Winter.—There is much more moonlight in winter than in summer. I do not mean merely that there are longer nights for the moon to shine in; but that in winter the brightest fortnight of the moon comes when the northern hemisphere of the earth is in the best condition for it, being turned toward the full moon, and so having it highest and longest above the horizon. You will see that in a minute, if you remember that the sun is lowest above the horizon in winter, because the north pole then leans away from him; and as the full moon is opposite to the sun, the north pole must lean toward the full moon when it leans away from the sun. Therefore each hemisphere has as much longer full-moonlight than darkness in winter as it has longer sunlight than darkness in summer. The nights of short moon in winter are also the nights of new moon, when there is the least moon to lose. And the contrary of all this holds in summer, when the moon is less wanted.

In consequence of the moon's orbit being 5° inclined to the ecliptic or sun's apparent orbit, the moon may be 5° higher above our horizon than the sun ever is ; or it may be so much lower. I remember a woman saying that the night when she saw a man stealing some fowls was the brightest she had ever seen. I had the curiosity to look whether there was any real ground for the remark, and was surprised to find there was ; for it happened to be the night of full moon at Christmas, and just at the time (which comes once in nineteen years) when the moon's ascending node coincided with the sun's ascending node ♈, and so the moon really was the highest that she ever is in latitude 52½°, viz : 38½° for the *colatitude* (or difference between 90° and latitude) + 23½° for the inclination of the equator to ecliptic + 5° for the moon's extreme elevation above the ecliptic, = 67°, or three quarters of the height from the horizon to the *zenith*, or highest point of the heavens.

ECLIPSES.

An eclipse of the moon is the moon passing through the shadow of the earth from the sun, which darkens the moon, as the shadow of a tree darkens the ground, only much more, because there is nothing to reflect and spread the light over the moon in shadow (subject to what is said at p. 116), as sunlight is spread by the air in some way that no other light is. You know that

the shadow of the tree stands still, and is equally visible
to you wherever you stand, if you are near enough to
see it at all. And so you never read in the almanac
there is going to be an eclipse of the moon visible in
Spain, or Norway, but not here; except that of course
it is only to be seen by the people on that side of the
earth which faces the moon just then.

But eclipses of the sun are always announced to be
visible at certain places, and astronomers travel into
distant countries to see remarkable eclipses of the sun ;
that is, either total eclipses or annular, which I have
described already (p. 91). If you put a small screen
between you and the fire, it will only hide the fire while
you are in one place ; that is, the eclipse of the fire is
only visible to you there, and if you move to another
part of the room you see no eclipse. So the new moon
may hide the sun from England while we can see him
here. And that is why there are more eclipses of the
moon than of the sun in any one country, though there
are in the whole world more eclipses of the sun.

But now comes the question why is there not an
eclipse of the sun and moon at every new and full
moon ? If the moon were always in the plane of the
orbit of the sun or earth (whichever we like to call it)
there would be, and that is why that plane is called
the ecliptic. But she is not: the sun and earth and
moon could not be represented by balls floating on the
same water. The moon's orbit is inclined to the ecliptic $5° 9'$; and therefore the moon is only in the ecliptic, or on a level with the earth and sun, twice in each
lunation, when she is just rising above or going below

the ecliptic at either of the two points called the nodes ; and therefore eclipses can only happen when the new or full moon is at or near a node.

Moreover, the nodes do not stay in the same place, but keep moving backward, so that each node is nearly 47′ less than half a circle or 180° from the previous one ; or they recede 19° 19′ a year, as I said at p. 101. This twisting motion of the plane of the moon's orbit backward is quite distinct from the revolution of the line of apsides of the elliptical orbit forward in about half the time. The recession of the nodes might be represented by dipping a sheet of tin into the water, inclined at 5° to it, and twisting it round from left to right, always keeping the same inclination ; while the advance of the apsides would be represented by an elliptical plate fixed loosely to the tin sheet by a pin through the focus, and turned from right to left, keeping the focus on the water.

If you put two candles close together, and hold a stick between them and the wall at a suitable distance, it will cast a narrow black shadow on the wall, bounded by two paler ones. Within the black shadow both candles are eclipsed, but in the pale shadow only one.* So the earth in a lunar eclipse casts a black shadow on the moon, called the *umbra*, within which the sun is totally eclipsed to the moon, surrounded by a *penumbra*

---

* This enables us to compare two candles ; for that one is the brightest which most illuminates the shadow of the other: therefore the right-hand candle is the best if the left shadow is the darkest. A lamp with a very wide flame would represent the sun more accurately, but two candles are easier to manage. When the sun is narrowed by an eclipse the fringes of shadows are narrower than usual on the corresponding sides of objects.

which is a pale shadow on the moon to us, and a partial eclipse of the sun to those parts of the moon which the penumbra covers. In like manner the moon hides the whole sun from the earth wherever the umbra or full shadow of the moon falls in a solar eclipse; and those parts of the earth which see the sun partially eclipsed would appear covered with a pale shadow to the people in the moon if there were any.

But the moon being much smaller than the sun, her umbra is a cone which runs to a point at 230,180 miles beyond the moon when the sun is nearest, and 238,010 when he is furthest off; and therefore falls short of the earth's surface if the new moon is near apogee, and reaches beyond it at perigee, and almost exactly to it at mean distances of the sun and moon (p. 92). When the umbra does not reach the earth it is only touched by the penumbra, and the parts of the earth which are swept over by the middle of the penumbra then see a partial but annular eclipse of the sun; for an annular eclipse is only a particular kind of partial one. Of course every eclipse begins and ends as a partial one. The places a little way off that middle line see a partial but not an annular eclipse.* If the moon is near enough for the umbra to reach the earth there is a total eclipse of the sun at all the places which it

---

* Some pictures of eclipses in books are likely to give an erroneous idea of an annular eclipse, by making the middle of the penumbra black like the umbra, which is contrary to the fact. No picture seems to me likely to make the matter clearer: in fact I had drawn one, and written a description of eclipses to suit it; but it was longer than this instead of shorter, and harder rather than easier, so far as I could judge. There is a good account of the principal 'solar eclipses' under that head in the English Cyclopedia.

covers. It can only cover a spot 147 miles in diameter, or move over a band of that width upon the earth, and the *totality* never lasts as much as 5 minutes at one place.

But the earth is so large that its umbra in a lunar eclipse reaches far beyond the moon, and may be as much as 5950 miles wide where the moon crosses it, and is never less than 5650. And as the moon is only 2165, there can never be anything like an annular eclipse of the moon. A total eclipse may last 1 h. 45 m., and it is an hour more than that from the first contact to the last, since the moon takes about an hour to move across her own width. The earth's penumbra is 10,200 miles wide at the mean distance of the moon. These things cannot be calculated without some geometry, so I only give the results. Those who wish and are able to do it must remember that the earth (or sun) is moving as well as the moon, and therefore we must take the synodical period of 708·734 hours in reckoning her velocity with reference to the earth's shadow. By that means the shadow may be treated as stationary; and the velocity of the moon is then 2234 miles an hour at perigee, where she goes fastest, and 2001 at apogee, and her mean velocity 2117; while her mean sidereal velocity belonging to her absolute period of 655·719 hours is 2288 miles an hour. The mean velocity per hour is of course the length of the orbit divided by the hours in which it is described; and the velocity at perigee or apogee × those distances = the mean velocity × the mean distance, by one of Kepler's laws which will be explained at p. 224.

The moon goes apparently, as well as really, east-ward through the shadow or over the sun, because she goes round the earth 13·37 times faster than the sun or the shadow, and the earth's rotation eastward * only makes them both appear to move westward equally.

The Moon is never Totally Eclipsed.—A lunar eclipse differs from a solar one in the boundaries of the umbra and penumbra being much less clearly marked. We either positively see or do not see the whole or part of the sun, and the part which is not eclipsed remains as bright as usual. For a solar eclipse is simplified by the moon having no atmosphere to refract the sun's rays as they pass by her. The only violation of the sharply defined boundary of light and darkness on the sun's disc is the unexplained phenomenon of *Baily's beads*, which are beads of light appearing between the edges of the sun and moon at the moment when an annular eclipse opens and closes. But the penumbra of the earth is only a gradually darkening shadow, into which the moon cannot be seen definitely to enter.

Moreover the umbra itself is invaded and colored red by rays which would pass by the earth and beyond the umbra altogether if they were not refracted and turned inward by our atmosphere, which they have to pass through as obliquely as possible in skirting the edges of the earth. And they are bent inward so much that they cover the whole umbra at the distance where the moon crosses it, leaving the undiluted umbra a shorter

---

* Appendix, Note XVI.

cone which does not reach the moon. In other words, there is no such thing as a total eclipse of the moon, or no time when she is quite invisible and does not receive some rays of the sun through the refraction of the earth's atmosphere; except that the rays are sometimes stopped instead of refracted by the earth's atmosphere being cloudy where they pass, and then the eclipsed moon is quite invisible, and stars passing behind her disappear suddenly without any visible cause.

Unless a new moon happens within $17°$ of a node, the sun escapes being eclipsed to any part of the earth; and unless a full moon happens within $11° 21'$ of a node, it clears the earth's shadow and is not eclipsed: passing through the penumbra only is not reckoned an eclipse of the moon. These limits are for mean distances, and are rather wider when the moon is nearer or the sun further. But the line of nodes must be crossed by the sun in his motion round the earth twice a year, and so there are at least two opportunities for an eclipse of each kind; and there may be more. There are never less than two eclipses of the sun every year, and there can easily be four partial ones. For there are $34°$ at each node within which one can happen; and if the moon gives the sun one partial eclipse $17°$ before reaching the node, she will be in time to give another at the next new moon, as the sun will only have moved about $29°$ in that time. Indeed there can be five solar eclipses, as there can be 13 new moons in a year, or 13 full moons, though not 13 of both. Therefore if there is an eclipse, lunar or solar, before January 11, there

may be a similar one at the end of December, if the position of the nodes is favorable, which it is the more
likely to be from their receding, so that the sun reaches
the same node again in 354·72 days, or very nearly 12
lunations (p. 108). A total or an annular eclipse must
be close to a node. There can be only three lunar
eclipses in the year, at intervals of 177 days or 6 lunations, as the moon can only be eclipsed once within the
22° at each node, and she may escape altogether. So
there are never more than 7 eclipses in a year, for if
there are 5 solar there cannot be 3 lunar, and *vice
versâ.*

CYCLES OF THE MOON, AND EASTER.

There are two cycles of the moon so nearly of the
same length that they are liable to be confounded,
though they are perfectly distinct. The first is the
Chaldæan **Saros,** or cycle of eclipses. If the sun,
moon, and one of the nodes are in conjunction to-day,
they will all three meet again in 18 common years and
10 or 11 days. For 223 lunations take 6585·32 days,
and 19 synodical periods of the sun and the moon's
nodes take 6586·28 days; which are not quite 20 minutes short of 18 common years and 10½ or 11¼ days
according as they happen to include 5 or 4 leap years.
Therefore after that the eclipses will recur in the same
order : not of course on the same days of the month,
but all 10 or 11 days later. There are generally 29
eclipses of the moon and 41 of the sun in that time.
But they recur, you see, nearly 8 hours later in the
day ; which may make all the difference of the eclipse

being visible or not, and may transfer it from one day to another. Therefore the Chaldæans made a more perfect cycle of three Saroses, which are 54 years and 31 or 32 days, within an hour.

The **Metonic Cycle** is so called because it was discovered by Meton, an Athenian, B.C. 433. It has nothing to do with eclipses, but is of far more importance than the Saros, as the time of Easter has been fixed by it as long as it has been fixed by any rule at all; for many centuries without any modification, and latterly with some; but still that cycle is the rule :—235 lunations take 6939·69 days, or so little less than 19 years of 365¼ days that they only differ by a day in 322 years. There is indeed a whole day's difference according as the 19 years include 4 or 5 leap years; and the complete cycle is 4 × 19 or 76 years. But the world has always been content to use only 19 *patterns* of years at once for finding the Easter moon : only they are now shifted in the calendar 7 times in 1200 years. Neglecting for the present that difference between leap and common years, which corrects itself every 4 years, and the moon's coming a day sooner in 322 years, we may say that the new and full moons come again on the same days of the same months every 19th year. Therefore there are only 19 out of the 30 days after the vernal equinox on which the equinoctial full moon can fall; and the 19 *golden numbers* prefixed to those days in the Prayer book calendar mean that the day against the golden number of the year is the day of Paschal moon, or full moon next before Easter Sunday.

For by the rule which has existed over all the world

since the first Council of Nice in 325, Easter is the Sunday after the full moon next after the 20th of March. The Council left the moon to be found as it might be; and further disputes arose on that; which were ended by Pope Hilarius in 463 ordaining what has ever since been the law of church and state, that the Paschal moon should not be the actual full moon to be found by astronomers (if it ever was so), but the 14th day of the moon by the Metonic cycle.

Consequently the Paschal moon often differs from the true equinoctial moon by a day or two, and Easter may be a week or even five weeks earlier or later than it would be if it followed the real moon. Indeed, unless an 'Easter meridian' were agreed on for the whole world, it might still differ five weeks in different places, even if it were fixed by astronomers and therefore made incalculable either forward or backward by anybody else. For if there is full moon in London very early in the morning of Saturday, March 21, Easter would be the next day there: but that same full moon may be on Friday night of March 20 at Exeter or Oxford by true time, and therefore would not be the Paschal moon there, which would be on Sunday, April 19, and Easter not till April 26—which it never is now.

The correction of the Metonic cycle for the purpose of keeping the ecclesiastical full moon tolerably near the real one on the average involves the whole subject of the reformation of the calendar, or the change from old to new style, which is intimately connected with astronomy. It occupies many pages in the books which treat of it fully, but I hope to explain all that is important of it in a short compass. The most complete treatise on it is Mr. De Morgan's in the 'Companion to the Almanac' for 1845 and 1846, and his article on Easter in the English Cyclopedia.* There is also a very clear paper about it in the Philosophical Transactions of 1750 by the Lord Macclesfield of that time, which prepared the way for the Act of Parliament, 24 G. II., cap. 23 (1751), for changing the style in September, 1752. But none of these contain a table for old style as well as new, nor do the old Prayer books, though they had one of a different kind for old style only.

THE CALENDAR.

It was known very early that a year is no exact number of days; and different nations had different plans for occasionally adding or *intercalating* days, to make up for the fraction lost in each year. It is not worth while to go into the history of these contrivances. The present scheme of three years of 365 days and a fourth of 366 was invented for Julius Cæsar by Sosigenes of Alexandria, B.C. 45, and lasted without alteration (except a temporary mistake corrected by Augustus) until the time of Pope Gregory XIII. It was then found that the real equinox fell 10 days before the nominal one of March 21, and that Easter had got four

---

* You will find there, and in Rees's Cyclopedia, a great deal also about the epact, or age of the (ecclesiastical) moon at the beginning of the year. But we can do as well without it, and the subject wants no superfluous complication. Mr. De Morgan also published a 'Book of Almanacs,' containing the 35 patterns of years for the 35 possible days of Easter, according to the rules: if they followed the real moon there would be 36 almanacs.

6

days wrong besides, from the error in the Metonic cycle. Ten days were accordingly struck out of the calendar between the 4th and 15th of October, 1582 ; and to prevent the error for the future it was decreed that every 100th year should lose its leap day except those divisible by 400.

This is the Gregorian or New Style, which was invented by Clavius, a Jesuit, and adopted in all the Roman Catholic countries before the end of 1582, and by the Protestant German states in 1700, but not by us till 1752, and Sweden the year after, and not yet by Russia and the Greek Church, which are therefore now 12 days wrong.   In 1751 Parliament enacted that the day after 2 September, 1752, should be September 14 : dropping thus 11 days, because we had got one day more wrong by allowing 1700 to be a leap year.   Of course due provision was made that nobody should lose or gain 11 days' interest on their debts ; nevertheless the Act caused riots among the common people, who cried out 'Give us back our eleven days,' as if they would die so much the sooner for the loss of them. Fortunately they were not masters.

Moreover you must remember in reading old books that the years legally began on March 25 until the change of style ; and therefore such an event as the execution of Charles I. on Tuesday, 30 January, 1649, as we call it now, is said in old books to have been in 1648.   A year of the average length of 365¼ days is called a Julian year ; but the years of the *Julian era* are reckoned from 1 January, 4713 B.C., an arbitrary epoch which was invented for reasons of no consequence now; so 1867 was the 6580th of the 'Julian era.' For there was no A.D. o, and the 19th century of our era (which probably begins 4 years after the real birth of Christ) began on 1 January, 1801, not 1800.

To turn old style into new add 11 to the day of the month in the century beginning 18 Feb., 1700; which becomes 1 March, as that February would have no leap day by N. S.: add 10 days in the *two* centuries from 19 Feb., 1500, 9 from 20 Feb., 1400, and so on.

Now let us see what amount of error the Gregorian calendar still leaves, after professing to correct itself every 400 years. A million Julian years are 365,250,-000 days; but a million equinoctial years are 365,242,-216 days. Therefore the problem is how to drop 7784 leap days in a million years in some neat and simple way: that is, one day in 128·47 years. But the Gregorian plan drops a day in 133·333 years in the long run. Therefore the error is 4·86 days in 128·47 × 133-·333 years, or a day in 3524 years. Accordingly Sir J. Herschel proposes to carry the correction a step further by making every 4000th year lose its leap day, though it is divisible by 400: which would go for about 28-000 years without an error of a day.

But I think the mere statement of the problem suggests the best solution of it: $128 = 32 \times 4$; and that is a number easy to remember, being formed by 7 successive duplications of 1, or 2^7. Therefore the simplest of all plans would be just to let every 128th year lose its leap day; and that only makes a day wrong in 35,-440 years; which is ten times more accurate than the Gregorian scheme, and seven times better than another which is commended by Sir J. Herschel, for dropping one day in 132 years (misprinted 128) by postponing the leap day of every 32nd year to the 33rd; which would also make the leap years no longer all divisible by 4.

Even if reckoning by centuries is thought essential, still the Gregorian scheme is not the best. It would be both more correct and more symmetrical to drop the leap day of every century except the fifth, instead of the fourth.* For that makes leap years of 2500, 3000, etc., instead of 2400, 2800, 3200, and only accumulates an error of a day in 4646 years: which again could be rectified in 5000 years more completely than the Gregorian error in 4000.

Correction for Easter.—The next question was how Easter was to be set right, and the Metonic cycle modified for the future, to prevent the ecclesiastical and real moons from getting wider and wider apart. For that purpose they wanted a new set of pattern years, to be indicated by the 19 golden numbers, and also a contrivance for shifting them again when they had got a whole day wrong permanently. The golden numbers of successive years have always run regularly from 1 to 19 without any dislocation. So the first thing was to put them against a new set of days between March 21 and April 18 (inclusive), and that arrangement was to last till the year 1700. Then they are to be shifted on this plan: at every century divisi-

---

* Friar Bacon in 1267 urged Pope Clement IV. to reform the calendar on a plan equivalent to this. He had found out somehow the length of the year more exactly than anybody else then or for a long time afterward. But he was put in prison for ten years instead, a fact which has been merged in the greater fame of Galileo.

ble by neither 3 nor 4 they are all to be advanced a day in the calendar, and to be set back a day in every century divisible both by 3 and 4, *i.e.*, by 12, and not altered in the others.

The reason of that contrivance was this. Suppose the sun and moon to be right at some century divisible by 12, which we may call 0 for this purpose. Then at the year 100 the Gregorian year loses a (leap) day, and so the moons of the next century will all come a day later than if that day had not been dropped, and the golden numbers must be advanced a day accordingly; and the same at 200. At 300 the year again loses a day; but by that time the Metonic cycle has put the nominal moon nearly a day too forward, and therefore ought to lose a day; and so they balance without any alteration. In 400 the year does not lose a day, and the moon does not, and so again they are right; and thus you may go on till 1200, which drops no day; but the moon has then lost a day since 900, and so the golden numbers have to be put back. But Clavius supposed the moon to lose a day by the Metonic cycle in 300 years more nearly than it does.

You may see the result of all this in the Prayer book in II and III of the ' General tables for finding the Sunday letter and the places of the golden numbers,' for as long as the world and the Gregorian rules may last. The small irregularity in the last three days of possible Paschal moons, which you will see in Table III, was introduced for an ecclesiastical reason not worth explaining here, and not affecting the principle of these calculations. The smaller numbers by the side of the centuries in Table II show at once how many days the golden numbers are advanced on the whole since 1600, which will evidently be 5 days in 1200 years on the average, being set back once in that time, and forward 6 times.

Now let us see how nearly this keeps Easter right, assuming it to have been set right in 1600. Though 235 lunations take 1h. 26m. less than 19 Julian years, they take 1h. 46m. *more* than 19 equinoctial years, so that the moons of every 19th year, or of every year with the same golden number, come a day later after 213½ years. In other words, if our years were correctly adjusted, by dropping two days in 257 years, a new or full moon of March 21 now would fall on March 22 in any year of the same golden number after 213 years, and on March 23 after 427 years, and so on. So that instead of the golden numbers advancing at the rate of 5 days in 1200 years, they ought to advance 5 days in 1067 years. Again 235 lunations exceed 19 Gregorian years (whose average length is 365·2425 days) so much that the moon advances a day in 227½ of those years, or 5 days in 1137 and not 1200 years: so that the Gregorian rule does not keep the moon right even for the average Gregorian year, which is itself wrong.

The result of all this is that the rules for keeping Easter neither keep it by the real equinoctial moon of each year, nor by a moon which is right on the average of a long period, either for the real equinox, or for the artificial year which we adopt, and sometimes leave it five weeks off the real time. Indeed it is very doubtful whether Easter is not *always* from 2 to 36 days later than the real anniversary of our Lord's resurrection.*

---

* See a full discussion of this point in Brown's *Ordo Sæclorum*.

It seems that even Clavius the Jesuit, who did the astronomical work of reforming the calendar for Gregory XIII., ventured to publish the suggestion that it would be better to make Easter always the Sunday after March 21 than to let all the great festivals and holidays except Christmas wander over five weeks of the calendar in the vain attempt to follow the moon. Perhaps before the year 2000 the world will be in a condition to revise the calendar and re-consider that question without prejudice.

To find the Days of the Week,—But whatever is done hereafter, Easter and the days of the week for given days of the month can only be found for the past and the present by the existing rules, and so it is important to understand them. The table in the Prayer book for finding the Sunday letter, or the day of the week for any given day of the month, is silent until 1600, and wrong from 1600 till September 1752. They copied the Gregorian tables into the Act of 1751 without taking the trouble to adapt them to this country. The proper table and rule are these:

0 A 1800.	For countries which	
1 G 14 Sep. 1752 to 1799.	changed their style	
2 F 2500.	in 1582 it must be,	
3 E 2400, 2300.	1 G 1700.	
4 D 2200.	2 F 1583 to 1699 :	
5 C every year of old style in all countries.	the rest as before.	
6 B 2000, 1900.		

Rule for Finding the Sunday Letter.—Add to the year

its fourth part, omitting fractions, and also the number set opposite to it or to the last century before it in
this table, and then divide by 7 : the remainder over
indicates the Sunday letter. But until the end of
February in leap years the letter above the one so indicated is the Sunday letter. When that is found you
have only to look at the Prayer book calendar, where
all the days marked with that letter are Sundays in
that year. The reason of the rule is that a common
year ends on the same day of the week as it begins, or
has one day over the 52 weeks, and a leap year two
days. Therefore starting from o, there are as many
days over beyond some entire weeks as the number of
the year + the number of leap years, and as many days
of the week over as the remainder after dividing those
days by 7. The arrangement of the letters in the calendar, beginning with A for January 1, is only arbitrary, and happens to require the constant addendum
of 5 to make the Sundays of all the years until the
change of style come right. After 1752 the dropping
of leap day in every century not divisible by 4 dis-
places the Sunday letters of the next 100 years by one
day.

To Find Easter.—Either of the tables in the Prayer
book for this purpose is right since 1752. I have made
a similar one for old style, which is right for every
country until it changed, and therefore here till 1753 ;
for Easter of 1752 went by old style. First come all
the possible days of the Paschal moon, and therefore
of the golden numbers, and a week more to reach a
Sunday when that moon is on Sunday, April 18. Then

come the Sunday letters of those days according to the calendar. The column headed O. S. has the golden numbers as they stood until the change of style. The next has them up to 1899 (inclusive) from 1753 in England, and from 1700 in Roman Catholic countries; and the following one from 1900 to 2199, unless the rule is altered before then. The column headed 1583 to 1699 is only for the countries which changed in 1582 (see English Cyc. Easter).

The way to use the table is this. Add 1 to the year and divide by 19, and the remainder is the golden number, no remainder corresponding to 19. Find that in the column which the year belongs to, and run your eye horizontally back to the day of the month opposite to it, which is the Paschal moon, or the 14th day of the moon according to the *epact* (p. 121, n.). The Sunday after it is found by the rule which I gave for the Sunday letter, and that is Easter Sunday.

6*

Possible days of Full Moon and Easter.		O. S.	1753 to 1899.	1900 to 2199.	1583 to 1699.		April. July. Jan. 7.
March 21	C	16	14	..	3	G	
22	D	5	3	14	..		
23	E	..	..	3	11	F	Sept. Dec. Jan. 6.
24	F	13	11	..	..		
25	G	2	..	11	19		
26	A	..	19	..	8		
27	B	10	8	19	..	E	June. Jan. 5.
28	C	..	..	8	16		
29	D	18	16	..	5		
30	E	7	5	16	..		
31	F	..	..	5	13	D	Feb. March. Nov.
April 1	G	15	13	..	2		
2	A	4	2	13	..		
3	B	..	..	2	10		
4	C	12	10	..	..	C	August. Jan. 3.
5	D	1	..	10	18		
6	E	..	18	..	7		
7	F	9	7	18	..		
8	G	..	..	7	15	B	May. Jan. 2.
9	A	17	15	..	4		
10	B	6	4	15	..		
11	C	..	..	4	12		
12	D	14	12	..	1		
13	E	3	1	12	..		
14	F	..	..	1	9	A	January. October.
15	G	11	9	..	..		
16	A	..	..	9	17		
17	B	19	17	17	6		
18	C	8	6	6	14		
19	D						
20	E	Old style everywhere.	1700 to 1899 Rome. 1753 to 1899 England.	English and Roman.	Roman only.		Sunday Letter of the first day of each month in the Calendar.
21	F						
22	G						
23	A						
24	B						
25	C						

THE TIDES.

Most people know as much about the tides as this, that they are somehow caused by the moon, and rise about 50 minutes later every day because the moon comes to the meridian so much later. But they are due to the sun also in exactly the same way, only not so much because of his greater distance. You know that every particle of the earth and water is attracted toward the sun with a force directly proportioned to his mass, and varying inversely as the square of his distance (p. 28). Therefore he attracts the water on the near side of the earth more, and that on the far side less, than the solid mass of the earth, which may be considered condensed at its own centre. And that comes to the same thing as if the earth's attraction both on the waters nearest the sun and farthest from him were a little diminished.

But that is not all. If you pull two balls not far apart with long strings of equal length held in one hand, you will also pull them toward each other, with a force which varies as the angle between the strings (so long as it is a small one), or as the distance of the balls apart divided by the length of the strings. In the same way the sun's attraction tends to squeeze in all the waters which lie at or near 90° from the point facing the sun, or to increase the earth's attraction on them. But this contractive force, or the *resolved part* of the sun's force on the sides of the earth, toward the centre, is only half the other separating or differential force, as I will show you presently. Therefore if you call the contractive force 1, there is altogether a force of 3 tending to make the water facing the sun, and at the back of the earth opposite to the sun, higher than the water 90° from those places.

The moon does the same in all respects, and in a greater degree; for although the general attraction of the moon on the earth is very small compared with the sun's, yet her differential attraction and her contractive force on the opposite sides of the earth are greater, because she is so much nearer, and both these forces depend on the proportion of the earth's radius to the distance of the sun and moon respectively. If you like to see the calculation of the actual amount and effect of the tidal forces of the sun and moon, and the proportion which they bear to gravity, or to the earth's attraction on its own water, it can be done as follows.

Calling the earth's radius and mass each 1 for shortness, the sun's distance is 23,064 (p. 62), and his mass 316,560. Therefore by the same kind of calculation as at p. 65, his attraction on the earth's centre is $\dfrac{316,560 \times \text{gravity}}{23,064^2}$; and for his attraction on the near and the far sides of the earth we must put 23,063 and 23,065 respectively for his distance. Then his differential force on the water at each of those places is the difference between his attraction there and at the centre; which you will find, if you take the trouble to work through the figures, in both cases very nearly $= \dfrac{316,560 \,\text{gravity} \times 2}{23,064^3}$. And as this 2 represents twice the earth's radius, the sun's differential force is to gravity as twice the earth's radius × sun's mass is to the cube of his distance. Similarly the moon's differential force is to gravity as twice the earth's radius × moon's mass ($\cdot$0123) is to the cube of her distance (60^3).

And to each of these half as much more has to be added for the contractive force; for that is the general attraction of the sun or moon on the earth × the small fraction which has the earth's radius or 1 for numerator and the sun's or moon's distance for denominator. Thus the same cubes of distance come in as before, but not the 2 in the numerator; and the whole tidal force of $\dfrac{\text{sun}}{\text{gravity}} = \dfrac{316,560 \times 3}{23,064^3}$; and of $\dfrac{\text{moon}}{\text{gravity}} = \dfrac{\cdot 0123 \times 3}{60^3}$.

You would find if you worked out these figures, that gravity is nearly 6 million times the moon's tidal force, and 13 million times the sun's, at mean distances; or the attraction of the earth on its own water is 4 million times greater, and the centrifugal force at the equator (p. 37) 13,800 times greater, than the tidal forces of the sun and moon together.

If the earth were a fluid sphere of nearly 21 million feet radius, and of uniform density = the present average density, the tidal force at every depth would be the same 4 millionth of the central attraction, since they both vary as the distance from the centre (p. 29). And as the weight of a prolate spheroid is to the sphere which it contains as their different axes, the tidal force would pull out the sphere into a spheroid whose semi-axis major exceeds the minor by $5\frac{1}{4}$ feet and lies in

the *line of syzygy*, pointing to the sun and moon. But
the inside density being five times that of the water
outside, attraction decreases inward less than the tidal
force, or proportionately increases ; and the result of
that and the solidity of the earth is that the tidal ellipticity due to sun and moon together is only a 6 millionth, or the highest tide is 3½ feet above the lowest
in the open sea.

You would find also that the moon's force on the near
side of the earth is rather more, and on the far side
rather less, than the fraction at page 133: in fact they
are about in the proportion of 21 to 20. There is no
such difference for the sun, because his distance is
23,064 times the earth's radius, instead of 60 times.

There is the same kind of excess in the force of the
earth on the near side of the moon ; but though the
earth is 81½ times heavier than the moon, yet the moon's
radius is only the 220th of the earth's distance, and the
excess of attraction on the near side of the moon over
the far side is only one 75th, and therefore quite inade-
quate to account for its supposed shape (p. 97), though
enough to have made her a slightly prolate spheroid
when she was fluid. The whole tidal force of the earth
on the moon is about 120 times that of the moon on the
earth, being greater as the earth's mass and gravity on
the surface exceed those of the moon, and less in the
proportion of their diameters.

If you cannot follow these calculations you may accept it as proved that the tidal forces of the sun and
moon are as their masses directly and the cubes of their
distances inversely. And their distances vary enough

to make a considerable difference in these proportions at different times. When the sun is at his nearest and the moon at her furthest, he is 360 times further off, and the cube of that is 46 millions ; but when she is nearest and he is furthest his distance is 410 times hers, of which the cube is 69 millions. So we have those two cubes in favor of the moon to set against 26 millions for the sun, who is so much the heaviest. Therefore the sun's tidal force varies from rather more than half to rather less than a third of the moon's.

Weighing the Moon by the Tides.—And on this an important result is founded. We have been assuming the moon's mass to be known and calculating the proportions of the tidal forces from it. But in fact it was just the contrary. The moon's mass was first ascertained from observations of the difference of the lunar and solar tides, *i.e.*, of spring and neap tides at various places (which I will explain presently), though it has since been done by other methods (see pp. 55, 248). Newton from imperfect measures of the tides made the earth 40 times as heavy as the moon, and Laplace 70 ; Mr. Airy called it 80 in 1856. Mr. Adams's figure is 81·5, making the moon ·0123 of the earth, as I put it just now, an easy figure to remember. Sir J. Herschel lowers the moon to one 88th of the earth ; and you will see some intermediate results at pp. 248, 250. The moon's mass is not affected by the late alteration of the mass and distance of the sun ; for the sun's tidal force remains the same as before, his mass and the cube of his distance being reduced equally.

The attraction of the earth then being diminished

in the line of centres of the earth and moon, and increased in all directions across that line (omitting the sun for the present), the water takes the form of a very slightly prolate spheroid pointing toward the moon: the elevation of high above low water, or the ellipticity of the spheroid, being just enough for the weight of the elevated water to balance the loss of ordinary gravity of the water by the tidal attraction. That is, it would be a prolate spheroid if the earth were not itself a far more oblate one, on which the tidal protuberance of only a few feet is superimposed; but in calculations about the tides the earth is always assumed to be spherical for simplicity. And the sun tends to form another spheroid pointing toward him.

But all this might take place and yet hardly any tide be visible, if the earth always kept the same face toward the moon, as she does to the earth. There would then be only a solar tide about a foot high, which would also move so slowly round the earth that its effects would be quite different from what we see now. The ebb and flow of the tide, by which alone it is felt as a great power over the world, depends upon the earth's rotation within the water, while part of it is held up by the tidal force. The easiest way to understand the effects of the rotation is to suppose the earth fixed, and the sun going round it from east to west in 24 hours, and the moon in 24h. 49m. The moon then drags the two opposite tidal waves after her at the rate of 1000 miles an hour at the equator, leaving the sun's action out of the question for the present.

Not that the water itself moves at anything like that

speed, or that much of it is carried round the earth at all, except in long periods. The thing that travels with the moon is the two alternate states of elevation and depression of the water at 90° apart. A wave is the transmission of a state, not of a body. The water is indeed moved to the very bottom of the sea, and a good deal of it moves forward, and some back again afterward, besides being lifted and let down again. Although the tidal wave travels westward with the relative motion of the moon, the tide itself moves toward an eastern as well as a western shore, because that is the necessary effect of the whole mass of water rising. And when the advancing water is stopped by land it can only dispose of itself by rising much higher than the 3 feet of the open sea. Waves raised by a wind stir the water to a very little depth, and not much water is carried forward in them. They 'break' on a shore because the friction of the ground stops the bottom of the water from going as fast as the top, which therefore tumbles over.

Neap and Spring Tides.—The sun has his two tidal waves as well as the moon, but of less than half the size ; and therefore it is best to consider the tide as mainly belonging to the moon and modified by the sun, as follows : At half moons, or *quadrature,* when the sun is 90° or 6 hours from the moon, they pull across each other, and the sun tries to make high water where the moon is making low water. The moon's tidal force being more than twice the strongest, prevails, but the tide is only due to the difference of the two forces, and so rises and falls least, and that is called *neap tide.*

When the moon is past quadrature and has not reached syzygy, or the line of new and full moon, the tide is kept in advance of it by the sun; but after syzygy the tide lags behind the moon, being kept back by the sun. Consequently the tide of any place is not regularly 49 minutes later every day, as if it obeyed the moon only, but sometimes as much as an hour later, and sometimes only 38 minutes. This is called the *priming* and *lagging* of the tides. But when the sun and moon are in syzygy, either in conjunction or opposition, they augment each other's tidal force and produce *spring tides*, which are the sum of the lunar and solar tides, and rise the highest and fall the lowest. And these again are greatest at the equinoxes, because then the sun is on the equator and the moon must be within 5° of it, and so they are in the best position for drawing the water from the sides to the front or back of the earth. For if you wanted to pull a sluggish globe round, you would wrap a string round it at the equator and pull in the plane of the equator.

The top of the tidal wave however does not really point to the moon at spring tides, but 45° or 3 hours behind it in the open sea, and much more where it is obstructed by land. For the inertia and friction of the water takes some time to overcome, and so the effect is always behind the cause. Spring tides are also a day or two after new and full moon, because the tidal force keeps accumulating for several days while the sun and moon are near together, and there is a greater amount of it in the four days with syzygy in the middle than in the four days before syzygy. So the hottest and coldest weather is after and not at the solstices. The more the tide is impeded by land the longer it naturally is behind the proper astronomical time: in London it is two days behind. Sometimes it has to come round islands, and is divided into two streams: consequently there are places where two tides come by roads of different lengths, and so rise and fall 4 times a day; and others where the low tide by one road neutralizes the high one by the other.

In running up gradually narrowing channels it rises much higher than on the sea shore; as high as 50 feet above low water at Bristol, and in some parts of the world 100, though it is only about 12 feet generally on an open shore. Sometimes the tide rolls up a river which gets gradually narrower, when the wind helps it, with a face like a wall and the velocity of a railway train, upsetting everything in its way. This is called the *bore* in the Severn and Avon and some other rivers, and the *eager* in the Humber, and it is far greater in some American and Asiatic rivers. On the other hand, when the tide has to make its way into a large sea through a narrow passage, like the Straits of Gibraltar into the Mediterranean, it is unable to produce any sensible rise and fall over such a sea.

But the tide sweeps rapidly over wide and level sands, so as to overtake and drown people sometimes, because a rise of a few inches then runs over a great area of sand; and it becomes soft under the water, like a bog, or 'quick,' because moving water lifts and carries sand and stones along with it, according to their smallness and (probably) the square of its velocity. For

the weight of the stones increases as the cube of their
diameter, but the surface only as the square, and the
power of the stream to move them varies directly as
their surface and inversely as their weight; and therefore varies inversely as their diameter, or as the cube
root of the weight, among stones of the same specific
gravity and general shape. A river goes on rising for
some time after 'slack water,' when things cease to
float upward, because the natural flow of the river
downward balances the tidal flow upward, but both
raise the water.

DISTURBANCES OF THE MOON.

There is scarcely one element of the orbits of the
planets or their moons that is not subject to continual
disturbance, by the attraction of every other body which
is large enough and near enough to affect them sensibly. All these disturbances in one way or other ultimately compensate themselves: some of them first
moving the body, or its orbit, a little in one direction,
and then an equal distance the other way: others producing recessions or advances of nodes or apsides,
which in time work round; and there is one remarkable acceleration of the moon, which has such an enormously long period that it may be said to increase
perpetually, though the time will come for it to change.
It is quite beyond the scope of an elementary book like
this to describe all the inequalities (as they are called)
of the moon alone, to say nothing of the planets.
Here and there I must notice a few of them, as I have
already precession and nutation—a disturbance upon

a disturbance of the earth. I will only select the most important of them for explanation, leaving you to pursue the inquiry in Sir J. Herschel's Astronomy, or Mr. Airy's Gravitation, which I think generally clearer in its treatment of this difficult subject; or in Newton's Principia, where the problem was first solved. One thing will be noticed here which was not known when those books were written.

Moon's Secular Acceleration.

—We saw at p. 43 that the minor axis of the earth's orbit has been long increasing; and it will increase for 24,000 years yet, while the major axis remains permanently unaltered. If so, the sun's average (not mean) distance from the earth and moon of course increases, and his power to disturb the moon decreases. Now let us see what that disturbance does. The sun attracts the new moon more than the earth, and the full moon less, because of their difference of distance : therefore at both syzygies the sun's differential force practically diminishes the earth's attraction on the moon. When they are equidistant from the sun he draws them closer together, as you would two separated balls by pulling them with strings of equal length. But this contracting force may be proved to be only half as great as the differential force, as in the similar case of the tides (p. 131). At intermediate places both forces are evidently less ; and at certain points nearer quadrature than syzygy they balance each other. Therefore on the whole the differential force greatly preponderates, and weakens the earth's attraction, and so enlarges the moon's orbit, and therefore her time of performing it is longer

than if there were no sun. But as his power of thus
retarding the moon slowly decreases with the increase
of his average distance, she is *comparatively* accelerated; and also comes gradually nearer to the earth—
about 4 inches a year.

As this retarding force is greatest in winter, when
the sun is nearest (p. 41), the moon falls most behind
her mean place in April, after half a year's excess of
retardation, and similarly advances 11′ 12″ before it
in October. This is called the annual equation; but
the mean place here referred to is the mean elliptical
place, which itself is found by applying the *equation
of the centre* (p. 41) to the mean place she would have
if she moved in a circle. Its greatest amount is 6° 18′,
which is equivalent to 12h. 24m. of time.

But now comes a most remarkable result of the latest investigation of this small advance of the moon of
only 12″ in a century; the meaning of which is that
the moon is 12″ past the line of syzygy at the time
when she would just be there if she had kept her mean
velocity of 100 years before. Or turning seconds of
space into time, as she moves almost exactly 1″ in 2
seconds, you may say that the full moon comes 24 seconds too soon at the end of every century. This was
first ascertained by Halley in 1693, from a comparison
of old observations, but not accounted for until Laplace
explained it, as I have described, nearly a century afterward. And all the astronomers who followed him
considered his calculations complete, though calculations of this kind are only approximate, and made
on the principle of taking into account all the quanti-

ties which are not too small to be appreciable. But Mr. Adams taking up the matter afresh in 1853 discovered that they had all disregarded something which turned out by no means too small to be appreciable when properly examined; for it was large enough to reduce the accelerating effect of the increase of the minor axis of the earth's orbit by one-half. Thus half of the known acceleration of the moon was again left unaccounted for; and where is it to come from?

That question has been answered mathematically by M. Delaunay, as it had been generally by Mayer and others before; and his explanation is now received even by our Astronomer Royal, who wrote an elaborate paper which he said proved Delaunay's three principal conclusions to be wrong; but he sent an '*addendum*' to it afterward, saying he found that they were right.[*] It is remarkable that Le Verrier, Hansen, and other foreign astronomers, altogether rejected for a good while Mr. Adams's correction of the lunar acceleration. But mathematical truth can afford to wait for recognition: it may lie a long time at the bottom of the well, but once brought up it never sinks again.

The theory now is that the earth's rotation is getting slower from the friction of the tides over the earth, and also from the drag between the moon and the earth and water in keeping up the tidal wave continually in a fresh place. For the earth turns eastward while the tides stay behind looking at the moon, and more water moves westward than eastward over the earth (p. 136); and all the friction of the water

---

[*] See Astronomical Society's notices for April, 1866.

moving westward uses up some force of the earth's rotation.

Even supposing no water to move over the ground, the mere elevation of the tidal wave does not run as freely as if the water had no internal friction and did not cling to the earth at all. An east wind sending waves along a field of corn transmits some force to the ground, and would move the field very slowly westward if it were afloat. And if the earth had had no original rotation, the moon dragging the tidal wave round it would by this time have given it some rotation.* Therefore the rotation eastward is diminished, and some of the force of rotation is used up in maintaining the tide.

This has been otherwise explained by saying that the tidal protuberance, like a mountain always standing out about 45° eastward of the moon (p. 138), serves as a lever or handle for the tidal attraction to take hold of and retard the earth's rotation, ' the water of the ocean being partly dragged over the earth as a brake.' But this either suggests the idea of the force and the work to be done being much greater than it is, or else leaves it unexplained how any drag at all is transmitted through the water to the earth, apart from the direct friction spoken of before. If the moon flew away the tides would go on running round the earth or rising and falling everywhere, for a time, but would gradually wear out; and the force we are considering is that which keeps them up. Moreover it must reciprocally act upon the moon as a tangential force, dimin-

---

* Appendix, Note XVII.

ishing her velocity and centrifugal force and radius of orbit, and therefore shortening her period a little, (p. 83).

Consequently the earth gets so much slower that any meridian, treated as a hand of the earth clock (and all our clocks only represent the earth's rotation), would be 12 seconds slow at the end of 100 years compared with a clock which had gone all that time at the rate of the earth's rotation 100 years before. If you wish to know how much a day the earth must lose to produce this result, you must remember that the daily loss accumulates by ' arithmetical progression ' into these 12 seconds in a century ; and that represents a loss of not quite the 66th of a second in the length of the day in 2500 years. So also you would find that a lunation is only half a second shorter than it was 2500 years ago.*

The real and apparent acceleration of the moon from both causes together has accumulated to $1\frac{1}{2}°$, or 3 times her own diameter, or 3 hours in time, in 2500 years. And that is quite enough to make a material difference in the places where a solar eclipse was visible, and therefore in some important dates of ancient history. For if the old accounts of a great battle say that it was stopped by an eclipse, and we can calculate that no great eclipse of the sun was visible there at any time but one, at all near the received date, we may be sure that was the time of the event. In this view, and also for finding the longitude at sea, as will be

---

* The sum of an ' arithmetical series ' is the number of its terms (here the days in 2500 years) + the square of that number, all × half the common difference (here the daily retardation).

explained hereafter, all the little disturbances of the moon are important, because her exact place can only be calculated by taking them into account.

Mr. Croll has pointed out another permanent effect of the tides on the moon herself. The solar tide wave must retard the motion of the earth round the centre of gravity of the earth and moon, in the same way as the lunar tide retards the motion of the earth round its own centre of gravity, and must therefore gradually diminish the distance of the moon. For if the earth turned in a month, the lunar tide would only be a stationary and therefore invisible elevation of the water in one place; but the solar wave would move round it in the month in consequence of the earth's monthly revolution round the joint centre of gravity; and that must destroy some of the force of that motion, or of the earth's centrifugal force round that c. g. The moon is not directly affected thereby, but the earth is brought nearer to the c. g., and therefore their distance is diminished, and their orbit round the joint c. g. made smaller and therefore quicker. I do not know that any calculation has been made of the amount of these disturbances, but it must be much less than the others which we have been considering.*

Measure of the Disturbing Forces.—We can easily calculate without mathematics the proportion which the differential and contractive forces bear to the ordinary earth-force on the moon, at the places where they are each greatest, *i.e.*, at syzygies and quadratures respectively. The earth's force on the moon is the

---

* See Mr. Croll's paper in the Philosophical Journal of August, 1866.

mass of earth + moon divided by the square of their distance, as we want to consider the earth at rest (p. 98); and we will take their mean distance for the unit of distance, to avoid large figures. The sun's distance from the earth will then be 383, and his distance from the new moon 382; and his attraction on them respectively $= \dfrac{\text{sun}}{383^2}$ and $\dfrac{\text{sun}}{382^2}$. The differential force is the difference of these; which by a common sum in fractions $= \dfrac{765 \times \text{sun}}{146{,}689 \times 145{,}924}$, which again you will find $= \dfrac{2\ \text{sun}}{(382{\cdot}5)^3}$. But the sun is 312,720 times as heavy as the earth + moon; and if you substitute that figure you will find the differential force $= \dfrac{1}{89{\cdot}5}$, or is that fraction of the earth's force on the moon, which force is 1 according to the assumptions we made. But the force at full moon is the difference between $\dfrac{\text{sun}}{383^2}$ and $\dfrac{\text{sun}}{384^2} = \dfrac{767\ \text{sun}}{383^2 \times 384^2} = \dfrac{2\ \text{sun}}{(383{\cdot}5)^3}$; which with the same figure for the sun's mass $= \dfrac{1}{90{\cdot}2}$.

You will see the consequence of this excess at new moon presently. But you have here the proof that the differential force varies inversely as the cube of the sun's distance very nearly, and directly as twice the difference of distance of earth and moon from the sun; for the 2 in the numerator represents that: all which is very like the case of the tides. Similarly we should find that the contractive force, which is half the differential when they are both at their maximum, is measured by the actual distance of the moon sideways from the line of syzygies, divided by the cube of the sun's distance. These last, observe, are general values of the forces, for all positions of the moon, which can easily be worked out (though at greater length) on the principle of the calculation which I gave just now.

It may seem odd, but it is the fact, that the magnitude of the disturbing forces on the moon depends on the proportion of the length of the year to her sidereal period of 27·32 days; which is 13·37. This cannot be proved here beyond observing that 13·37^2 or 179 is twice the denominator of the fraction which represents the proportion of the sun's disturbing force to the earth's attraction (p. 147). And this, like the tidal force, is not affected by the late correction of the sun's mass; for the mass and the cube of the distance had to be altered equally, as the length of the year depends on both of them. Moreover the moon's period is lengthened by an eighth of that fraction, or a 716th; as it would be if the earth's mass were reduced a 358th. The mean distance is also increased a 1432nd, for a reason which you will see at p. 224. It is convenient to mention here, that where a quantity is increased by a small fraction, its square and cube practically increase by twice and thrice that fraction, and its square and cube roots by half and å third of it, *i. e.*, by doubling or trebling its denominator.

Tangential and Radial Forces.—Now let us see what the differential and contractive forces do, besides producing that long acceleration of the moon which I have spoken of, and which is only a very small part of their effects. As both forces are acting at every part of the orbit except syzygy and quadrature, where they alternately vanish, they must combine to produce a resultant force in some direction between them, as two winds blowing across each other would send a ship or a ball in some diagonal course between them. As the differential force always acts from the line of quadratures, QQ in the figure at p. 153, and is proportional to twice the moon's distance therefrom, it always accelerates her from quadrature to syzygy and retards her from syzygy to quadrature. And as the contractive force always acts toward the line of syzygies SS, and is proportional to the moon's distance therefrom, it also accelerates her toward syzygy and retards her after syzygy. Thus a part of both these forces, whenever they *both* exist, is always resolved into a tangential force, which accelerates before syzygy and retards after it; and the rest is resolved into a radial force, which acts with the earth's attraction for $35°$ on each side of quadrature, and more strongly against it for $55°$ on each side of syzygy. At $55°$ from syzygy they balance each other, and the radial force vanishes; but the tangential force is greatest half way between syzygy and quadrature, at the places called *octants;* and there it amounts to $\frac{3}{4}$ of the differential force at the adjacent syzygy: but this cannot be proved here.

You will easily see that the sun is really a little further from the moon than the earth at true quadratures or $90°$ from syzygy. But the difference is only $4\frac{1}{2}'$, the angle corresponding to half the moon's distance

divided by the sun's; which is much too small to affect
any calculations that can be given here; and so is the
inequality arising from the sun's force in the plane of
the moon's orbit being rather less when she is not also
in the ecliptic—as she never is quite, except at the
nodes. In the figure at p. 153 I have marked the forces with arrows according to their directions, and I
have given the radial force at syzygies two arrows,
because it is double of that at quadratures, except so
far as they all vary with the moon's distance from the
earth: *e.g.*, the tangential force at Λ (say) 50° after S_2
exceeds the tangential force at P 50° after S_1 as much
as EA exceeds EP.

Variation.—This constant acceleration up to syzygy
and retardation after it, makes the moon alternately 35′
42″, or rather more than her own width, before and
behind her mean longitude; and this is called her *variation*. You would probably expect it to carry her
further away from the earth at syzygy than quadrature.
But it does just the contrary. For the moon going
fastest at syzygy, from a cause different from the earth's
attraction, is least drawn out of her forward course by
the earth and goes further on toward quadrature; and
so the orbit becomes an oval with its sides at syzygy
and its ends at quadrature, and the minor axis a 70th
less than the major (supposing the undisturbed orbit to
be a circle). But you must not confound this secondary
oval with the much more elliptical general orbit of the
moon, of which this is only a small disturbance.

Parallactic Inequality.—We saw just now that both
the differential and contractive forces, and therefore

their resultant tangential force, are a little greater on
the near side of the orbit than on the far side. Consequently the 'variation' at new moon exceeds that at
full moon by the small amount of 2' 6''. If you work
out the calculation at p. 147 completely, keeping the
moon's distance as 1, but increasing the sun's to 400,
as it used to be reckoned, and increasing his mass to
357,050 in proportion to the cube of the distance, you
will find the average disturbing force the same as before, but the difference between its two extremes a little less. And with some trouble you might find that
this difference of the differential force $= 6 \times$ sun's mass
(keeping that the same) divided by his distance⁴ (or
distance × cube of distance). But since the mass bears
a fixed proportion to the distance³, that leaves this inequality to vary inversely as the sun's distance, as his
parallax does. Hence it is called the parallactic
inequality; and this alone of all the disturbances
gives any measure of the sun's distance. In fact its
observed excess over the amount due to the sun's old
distance raised the first suspicion in 1854 of that distance being too great, as I said at p. 85.

Subject to this small difference of 2', the 'variation'
compensates itself in opposite halves of the orbit—provided the two halves are alike in the long run. But
they are not : for there is a gradual decrease in the
length of the radius vector, as explained at p. 142 ;
and Professor Adams found that the consequence is
that the 'variation' does not quite correct itself every
fortnight on the average, as Laplace and everybody
else had concluded that it did ; and that it produces

a secular retardation about half as great as the acceleration due to the other cause, as I have already said.

The Advance of the Apsides may be shortly proved as
follows, though it requires a long investigation to trace
all its causes through all the positions of the moon and
of her orbit. It is only necessary to remember first,
that the sun's disturbing force on the whole weakens
the attraction toward the earth.

The undisturbed apogee is the place where the moon
would begin to move toward the earth if there were
no sun; but if the earth's attraction is weakened there,
it cannot pull the moon round the corner so quickly,
and she will carry the apse along with her a little.
At perigee she begins to leave the earth again; but
if the earth's attraction is weakened there, she will
begin to leave sooner than she would otherwise; or
that apse comes sooner, or recedes. But these opposite
effects by no means balance each other; for the differential force varies as the moon's distance from the
earth, which is about a 19th greater at apogee and a
19th less at perigee than at mean distance. Besides
that, the earth's attraction is itself about a 9th less at
apogee and a 9th greater at perigee than at mean distance: and we found at p. 147 that the mean differential force is about a 90th of the earth's mean attraction. Therefore the actual differential force will
diminish the earth's actual attraction at apogee by
$\frac{20}{19} \times \frac{10}{9} \times \frac{1}{90} = \frac{1}{82}$, but at perigee only $\frac{18}{19} \times \frac{8}{9} \times$
$\frac{1}{90} = \frac{1}{107}$ whenever the apses are in syzygy and the

effect of the forces in disturbing them is greatest. The effects of the contractive force must be opposite to those of the differential force ; but as I said at pp. 148, 149, only half as great, and lasting much less time. Therefore on the whole the advance of the apses preponderates.

The sun then would drive the apses forward even if he stood still ; for we have said nothing yet about his motion. But he does go round the earth the same way as the apses, and therefore drives them faster. He also stays in company with a progressing apse, keeping up its progress, longer than with a receding apse which he only meets, and thus makes them progress still more. By these two causes the advance of the apses is made twice as great as it would be without them. In the same way the apses of the earth's orbit are carried round by the attraction of the planets, the greatest weight of them by far being outside the earth's orbit ; but in 110,880 years instead of 9, as their disturbing force is much weaker than the sun's upon the moon.

Change of Eccentricity.—The sun disturbs the eccentricity of the moon's orbit so much that it is worth while to attempt the explanation of that also. In this figure, which I have described already, P and A are perigee and apogee for this one position of the orbit, or of syzygy and quadrature with respect to perigee and apogee: other

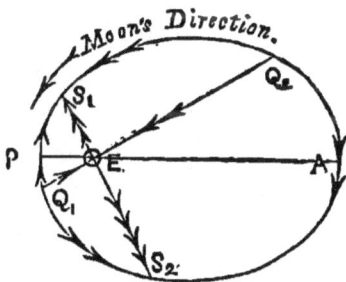

positions would want other figures, but this will serve our purpose.

1. When the moon is at the syzygy S_1 she is approaching perigee, or the radius vector is decreasing; but the radial force there acts against it and tends to keep the radius from shortening so much, and therefore makes the orbit less eccentric (p. 40). At the opposite syzygy S_2 the radius is lengthening, and the radical force tends to lengthen it still more, or to increase the eccentricity. We must see then which prevails. At S_2 the moon is further off, and going slower; and so the disturbing force is both greater and has a longer time to act (p. 153): therefore the increase of eccentricity prevails thus far.

2. At Q_1 the radius is lengthening, but the contractive radial force acts against it, and therefore diminishes eccentricity. At Q_2, after apogee, the radius is shortening, and the contractive force helps to shorten it and therefore increases eccentricity. And Q_2E is greater than Q_1E, and therefore again the increase prevails. At these places there is no tangential force.

3. The tangential force, retarding from syzygy to quadrature, diminishes the moon's velocity at P, and therefore diminishes her centrifugal force, or power to fly further off from perigee, or to increase her radius vector; and so the eccentricity is diminished by the tangential force at P. But it is increased by the tangential force retarding the moon at A and weakening her power to resist the earth's attraction, which shortens her radius faster than if she moved quicker and had more centrifugal force there. And the force is

both greater and acts longer at apogee than perigee, as before; so again the eccentricity is increased. At intermediate places near Q_2 and S_2 the effects of the tangential force balance each other pretty nearly.

The result is that all the disturbing forces increase the eccentricity when the moon has to pass through the apses before quadrature and after syzygy. You may easily infer that the eccentricity is diminished when she has to pass the apses after quadrature and before syzygy. And when they lie in either syzygy or quadrature the forces balance each other and do not disturb the eccentricity. Nevertheless it is greatest when the apses are in syzygy and least when they are in quadrature. For those places move round the earth with the sun in a year, while the apses take nearly 9 years to revolve in the same direction; therefore S_1 is approaching P whenever they are in the position of this figure, in which the eccentricity is increasing; and it goes on increasing till syzygy has reached the apse, and consequently it is greatest then. Similarly it is least with the apses in quadrature. And on the whole it varies so much as to be half as great again with the major axis in syzygy as in quadrature.

Evection.—The variation of eccentricity and the irregular motion of the apses produce together the largest and earliest observed of all the displacements of the moon; which was called Evection, or the carrying away of the moon from her mean longitude (not her mean anomaly but her mean elliptical place) by as much as $1°\ 20'$ alternately backward and forward; making her oscillate through 5 times her own width

7*

in the time of the sun's passing perigee twice, or about
a year and six weeks. Consequently it depends on, or
is a *function* of, the difference of longitude of sun and
moon, and also of the true anomaly or distance from
perigee, on which the motion of the apses depends.

Recession of the Nodes.—In all these cases of disturbance the orbit that we speak of as if it were a ring
capable of being moved, is the *instantaneous ellipse,*
which the moon would go on describing thereafter if
the disturbances were stopped. The recession of the
moon's nodes is caused by the attraction of the sun on
the moon, exactly as the precession of the equinoxes, or
nodes of the equator and ecliptic, is by the attraction
of the sun and moon on the equatorial protuberance of the earth, which may be considered a ring of
satellites stuck together. We need only consider the
motion of one satellite to explain them both.

But the effect of the sun's and moon's attraction on
the equatorial ring or ' elliptical excess ' is much less
than it would be if there were not a sphere many times
heavier inside, which has to be dragged round with it
in giving the twisting motion to the earth's axis. The
bulk of the elliptical excess is a 149th of the whole
earth (an oblate spheroid and the sphere within it being
in the proportion of the squares of their different axes) :
but the outside is not half as dense as the inside (p. 26) ;
and therefore the mass of the whole earth is probably
300 times that of the elliptical excess. Again that is
not really concentrated into a ring, but spread over the
whole surface, from 13 miles thick at the equator down
to nothing at the poles. Moreover the forces which

produce all the disturbances vary as the distance between the two attracted bodies, and the moon is 60 times farther from the earth's centre than the equator is. From all these causes together, and the absence of the moon of course in disturbing her own nodes, the earth's nodes or equinoctial points recede 1390 times slower than the moon's.

1. As the sun occupies all sorts of positions with respect to the nodes during a year (or rather less, since the nodes revolve backward in nearly 19 years) we must consider them in succession. First let us take the nodes in quadrature. Then the tangential force urges the moon forward as she rises to syzygy and her greatest latitude from the ecliptic. Besides that, there is always a force toward the ecliptic, as the sun's differential force is always trying to pull the near side and push the far side of the moon's orbit down to the ecliptic, except when he is himself in the line of nodes and therefore in the plane of the moon's orbit. This is called the *resolved force* of the sun toward the ecliptic.

Now if the moon's apparent path in the heavens, rising from the ecliptic and coming down to it again, is opened out into a flat picture, it will look like the path of a stone or a ball shot from the ground at an angle of 5°, and coming down to it again at the same angle, the place of falling corresponding to the next node. But that disturbing force of the sun which acts toward the ecliptic is the same as if a wind blew down upon the ball; and the effect of that would manifestly be to make it reach the ground sooner, or the node to recede. Also while the ball is rising, the downward force would evidently keep diminishing the angle of inclination of its course ; but would increase it while falling ; and so the inclination would end as it began ; though the moon or the ball has not risen so high above the ecliptic or the earth, nor gone so far, as if the disturbing force had not acted.

The tangential force, which always urges the moon forward from quadrature to syzygy, tends to postpone her arrival at the next node, and also to make her course flatter, or the inclination of the orbit less ; but the same force acts the contrary way from syzygy to the next quadrature and node, and so those two balance each other.

2. Next let the nodes be in the line of syzygy. As the sun is then in the plane of the moon's orbit, he plainly can do nothing toward pulling or pushing the moon out of her orbit, or altering either the nodes or the inclination. And from these two cardinal positions of the nodes we may conclude, that as they recede through the whole lunation in one case, and never advance in the other, they must on the whole recede, even if they advance a little in some intermediate position.

3. But we may as well complete the inquiry by seeing what happens when they are neither at quadrature nor syzygy ; and first, let each node be after quadrature. Then as the moon comes down to node from quadrature, toward the sun, he pulls her forward out of the course she is taking, and so makes the node advance, *i. e.*, makes her reach the ecliptic later. In

the opposite quarter of the orbit, the force acts similarly, and there also makes the node advance. But in the rest of the orbit the disturbing force is toward the ecliptic as before, and therefore makes the nodes recede.

4. Lastly, let the nodes be before quadrature and after syzygy. Then while the moon goes up from node to quadrature, leaving the sun, he also pulls her back, which makes her course less parallel to the ecliptic, as a head wind would make the course of a ball, while rising, still less parallel to the earth : and that postpones her arrival at the next node, or makes it advance. And the same thing happens in the opposite quarter, as usual. But in the rest of the orbit the force is downward, or toward the ecliptic, and so the nodes recede. Therefore on the whole, as the nodes cannot advance through more than two quarters of the orbit in any lunation, or through more than half that quantity on the average, and recede in all the rest, our former conclusion was right, that the recession greatly preponderates.

If we followed the inclination also through the last two cases, we should find that it is diminished when the nodes are in the third position, but increased when they are in the fourth, and therefore is not altered in a complete set of lunations, when the nodes and sun have gone all round each other, except by the minor disturbances beyond the scope of this book.

Two very small disturbances of the moon have been discovered by Professor Hansen, both due to Venus. We shall see at p. 167 that she retards the earth, and therefore enlarges its orbit or distance from the sun,

for 120 years (which Mr. Airy discovered), and therefore increases the moon's secular acceleration for that
time ; and then diminishes it for another 120 years :
the effects accumulating to 23″ in that time.   The
other is of a more complicated kind, and accumulates
to 27″ in 136 years.   Sir J. Herschel says these alone
were wanting to account for all the observed inequalities of the moon's motion, of which there are more
than 50 altogether.   The first idea of the 'lunar
theory,' or of the moon's disturbances being chiefly
caused by the sun, was conceived by Jeremiah Hor-
rocks, who died at 21, about a year after the transit of
Venus in 1639, which he alone predicted.

CHAPTER IV.

THE PLANETS.

THE earth is by no means the only body that goes round the sun. In the earliest times of astronomy it was observed that there were five stars unlike all the rest in their behavior: apparently going round the earth (independently of their daily rising and setting) in longish periods, though with some irregular motions backward and forward, two of them taking less than a year to go round, and the other three taking nearly 2, 12, and 30 years. These five *wandering* stars neither kept the same distance from each other nor from the other stars, which are called fixed because they do not visibly change their places, except to the very small amount which I shall describe hereafter; and they were therefore called *planets*. The ancients either named them after some of their gods, or their gods after them: for it is by no means agreed which came first, the heathen *mythology* and worship of false gods, or the belief in the planets influencing the bodies and fortunes of men: the study of which is called *astrology*. I have no doubt that came first, for reasons which are not material to state here. The planets have still kept these old names, and no doubt always will. For it may be observed that no civilized nations, nor perhaps any nations now, can invent 'proper names,' *i. e.*, names for persons or places: they can only copy or compound old ones.

The names of those five old planets are Mercury,
Venus, Mars, Jupiter, and Saturn ; and the days of the
week are still named after them, with the addition of
the sun and moon: either directly, as Saturday, Sunday, Monday, or through the Saxon names for the
others: thus Tuesday is the day of Mars (Tuisco),
Wednesday of Mercury (Woden), Thursday of Jupiter
(Thor), and Friday of Venus (Friga).

These five planets with the sun and moon were also
the principal characters in that ' host of heaven ' which
the idolaters of old worshipped long before the Greeks;
some of whom, you remember, took Paul and Barna-
bas for Jupiter and Mercury, and others worshipped
an image of Diana, the goddess of the moon.   The first
Greek historian Herodotus says that the first Greek
poet Homer, who probably lived about the time of the
prophet Elisha, borrowed the names of Jupiter and
most of the other Grecian gods from the Egyptians,
who also practiced astrology or divination by the planets, as the Chaldæans did (Herod. II., 4, 50, 53, 82).
Baal and Ashtoreth were the gods and idols of the Sun
and Venus, and Nisroch probably of Saturn.   He is
represented in Assyrian sculptures encircled by a ring,
which was the symbol of Time, the Greek name of
Saturn ; also as a human eye, which resembles the oblique view of Saturn and his Ring.   They represented
Venus with a crescent (p. 202), which can no more be
seen without a telescope than Saturn's Ring * (see Mr.
Proctor's Saturn, p. 197).

Indeed no one who has inquired into both subjects

---

* Appendix, Note XVIII.

can doubt that Pagan idolatry was closely connected with the belief in the influence not only of the sun, but of the moon and planets, on the bodies and affairs of men.* There is reason to believe that the things translated *groves* in some passages of Scripture, which were 'built' and 'set up on every high hill and under every green tree,' and carried out of the house of the Lord by the good king Josiah and burnt, and therefore certainly not groves of trees, were wooden machines representing the planets and their apparent motions,† and used as images of the powers then supposed to rule the world, rather than the Lord who made the heavens and all the host of them, and will one day 'make new heavens and a new earth wherein dwelleth righteousness.'

When Copernicus found out that the earth goes round the sun, he found the same of the planets also ; in other words, he discovered that the earth is one of the planets. And as I said before, people were at last driven to accept his theory or explanation of the planets' motions by finding that no other would account for them ; for the planets would not appear where they ought according to any of the other theories. When we come to the aberration of light (p. 211) you will see a direct proof of the earth's motion in the apparent annual motion of all the stars, though that was not observed till long after Copernicus.

It was long before the reason of their motions was discovered, or the full number of the planets. Sir Isaac

---

* See Faber's 'Origin of Pagan Idolatry,' etc.
† See Landseer's 'Sabæan Researches.'

Newton knew of no more than the five old ones and the earth, because the telescopes of his time were not large enough, that is, did not take in enough planet-light at a mouthful, to show the smaller planets and the more distant ones which have been discovered within the time of people now living. Copernicus did not even get so far as to discover that the planets describe ellipses, although it was known to Hipparchus 1700 years before that the sun is not always at the same distance from the earth, because his disc is larger at some times than at others. The elliptic motion was discovered by Kepler soon after the year 1600, together with two other remarkable laws of planetary motion, which I will explain hereafter. But he, like Copernicus, only found that the planets observed these laws of motion as a fact. Newton found a reason for them, and proved that every planet round the sun, and every moon or satellite round a planet, must observe them.

The dimensions, weights, and motions of the whole solar system, which means the sun with its planets and their moons, are now considered to be as follows:

1. The first planet is **Mercury** ☿, 35 million * miles from the sun at mean distance, and going round him in nearly 88 days. By 'days' I mean our days, and not the planet's own days; for if we want to compare their periods or times of going round the sun, we must measure them all by days of the same length; and you

---

* I give the more precise figures in the table generally. The distances in proportion to the earth's mean distance are independent of the accuracy of that, and of course the periods; and so are the masses in proportion to the sun, but not to the earth.

will soon see that there is no known relation whatever between a planet's period or year and the length of his day or time of rotation.

His diameter must now be called 3050 miles; and so he is not quite 3 times as large as the moon, but more than five times as heavy, because he is rather denser than the earth (1·115), which is 15·4 times as heavy and 17·5 times as large as Mercury. The sun is nearly 5 million times as heavy. But though Mercury is so small, he turns on his axis slower than the earth, his sidereal day being 5½ minutes longer than ours. He is always so close to the sun, never more than 29° off, that he is difficult to observe accurately, and can never be seen except as a 'morning or evening star,' just before sunrise or after sunset.

Mercury's orbit is the most inclined to the ecliptic of any (7°) and also much the most elliptical, the eccentricity being ·2056 ; and therefore his greatest distance is ·4666 and his least ·3075 times our mean distance. But even this great eccentricity only makes the axis minor about one 50th less than the major (see p. 39). I said that the earth moves through space 65,500 miles an hour ; but Mercury goes much faster, viz., 105,000, or nearly 30 miles a second on the average, but 35 at perihelion which is 28 million miles from the sun, and 23 at aphelion which is 42 millions, calculated as I explained for the moon at p. 115.

The apparent *size* of the sun's disc to a planet, and the light and heat received there, vary inversely as the square of the distance, but the apparent *diameter* of the disc varies inversely as the distance only. And as

the sun's apparent diameter here is 32′, you will easily
calculate that at Mercury it varies from 1°38′ to 2°29′,
and that the apparent size of the sun, and the light and
heat, are from 5 to nearly 11 times as much as they are
here. The apparent diameter of Mercury to us of
course varies still more, viz., from 5″ when he is
beyond the sun, to 12″ when he is between the sun
and us. The eccentricities of the other planets' orbits
are so small that I shall not notice these distinctions
between their greatest and least distances.

2. Venus ♀, the next of the planets, has the most
circular orbit of them all, its eccentricity being only
·007, and the semi-axes minor only 1600 miles less than
the major; which is 66 million miles, or ·7233 of our
distance from the sun. Her period is 224·7 days, and
her sidereal day 39 minutes less than ours. The orbit
is inclined to the ecliptic 3° 23′. Consequently she
travels nearly 77,000 miles an hour, and the apparent
diameter of the sun there is 44′, or half as wide again
as he appears to us. The diameter of Venus is 7770
miles, or a little less than the earth's; but she weighs
one-fifth less than the earth, and is $\frac{1}{401840}$ as heavy as
the sun, her density being only ·836 of the earth's.
She gets twice as much light and heat as we do; and
in fact her brightness prevents her from being as well
observed as some of the more distant planets. It has
been ascertained lately by experiments that Venus is
ten times as bright as the brightest part of the full
moon; which is probably owing to the atmosphere of
Venus being a much better reflector than the rough
surface of the moon without an atmosphere (p. 93).

From that cause we have not yet ascertained how much her axis leans, or the amount of her spheroidicity. Probably it is much the same as the earth's; but the axis is thought to lean very much more. Gravity at her surface is less in the same proportion as her mass, the distance of the surface from the centre being practically the same as here. Her apparent diameter is as little as $9\frac{1}{2}''$ when she is on the other side of the sun, and as much as $61''$ when she is nearest to the earth; and she exhibits phases like the moon (see p. 106).

Venus, like Mercury, never appears far from the sun : 47° is the greatest angle, or apparent distance, called the *elongation*, ever made by the lines of sight from us to the sun and Venus. Consequently she never appears but as a morning or an evening star, rising a little before or setting a little after the sun; but she is sometimes visible by day, even without a telescope. Her transits over the sun will be spoken of hereafter. Thirteen years of Venus agree with 8 of the earth within a day, and in that time they have 5 conjunctions, or 5 synodical periods; which produces that disturbance of both their orbits which I referred to at page 159, accelerating the earth and retarding Venus a little for 120 years and then reversing the effects for 120 years more, as I will explain for the much larger inequality of the same kind between Jupiter and Saturn.

3. The next planet is the Earth ⊕, of which I have said enough in Chapter I.

4. The first planet beyond the earth, or the first of what are called the *superior* planets, is Mars ♂. It would have been better to call them *exterior*, leaving

the term 'superior' for the much larger but less dense
ones which come after Mars, and after a great gap in
the system, and rotate more than twice as fast notwithstanding their size, as you will see presently.   His
mean distance from the sun is 139 million miles, or
1·5237 of the earth's distance, and he performs his
circuit in 687 days, or a little less than two years : and
therefore his velocity is about 53,000 miles an hour,
and the sun's apparent diameter there is 21'.   The
eccentricity of his orbit is one 11th, and it is inclined
1°·51' to the ecliptic.   He again is small, his diameter
being only 4155 miles; and his density must be only
·65 of the earth's or a little more than the moon's, as
the earth is 8½ times as heavy, but only 5½ times as
large; and the sun is 2,680,337 times as heavy.

Mars gives us better opportunities of seeing him
than any of the other planets: better than his supe-
riors, because the nearest of them is never less than 8
times as far off as he is sometimes; and better than
the two inferiors, because they are too near the sun.
You will see by adding and subtracting our solar distance to and from his, that he is at one time only 48
million miles from us, and at another 230.   Consequently his diameter appears nearly 5 times larger,
and his whole disc 25 times larger, at one time than
the other, and his apparent diameter varies from 4″ to
18″.   Astronomers have been able to observe, what
they have not in Venus or Mercury, the inclination of
his equator to his orbit, and find it 28° 42', or not
much more than ours, and its inclination to our ecliptic 3° 18'.   His day is as much longer than ours as

Venus's is shorter, 39 minutes, or as other authorities say, 37.

There are appearances of snow at his poles, which decreases in their summer; and there is something in his composition which makes him generally look red; but some parts look green, which are therefore thought to be water or vegetation. There are clear indications of an atmosphere in all the planets, though few, if any, in the moon. His heat from the sun is less than half of ours, and gravity on his surface about the same as on Mercury, and less than half of what it is here.

The spheroidicity of Mars is given in most books as a 60th, or 5 times greater than the earth's; and Mr. Main's Greenwich observations in 1862 made it as much as a 38th. But even the smaller of these amounts is hardly possible. The oblateness does not depend on the size, for the reason given in the note to p. 9, but varies as the square of the velocity of rotation, and inversely as the density or attraction of the globe on its own matter. As Newton divined long before anybody proved it, the oblateness is proportional to the centrifugal force at the equator divided by the equatorial attraction of the globe (p. 241). It also decreases (down to a certain limit) with an increase of density toward the centre, like the tidal ellipticity (p. 134). If the earth's density were uniform, its ellipticity would be a 230th instead of a 298th.*

---

* I believe the most complete treatise on the difficult problem of ' the figure of the earth ' is Archdeacon Pratt's. And in the Phil. Mag. for Feb., 1867, he comes to the conclusion that the polar diameter is 7899·74 miles, and that of the equator (which he does not believe to be elliptical) 7926·6, and therefore the ellipticity a 296th, and the diameter of an equal sphere 7917·6 (see p. 24).

8

Therefore Mars having ·65 of the earth's average density, and rather less velocity of rotation, ought to have an ellipticity of a 205th if its density varies like the earth's; which could hardly be seen. In order that such a globe may be 8 times more oblate than the earth, either its density must be very much greater outside than inside, which is almost impossible; or it must have turned very much faster when it was fluid than it does now (which again is very unlikely, see p. 234), and have become solid in that shape before it began to turn slower; or else it must be much denser at the poles than at the equator for the matter at the poles to balance the greater bulk at the equator: which is equally unlikely. And on any of those suppositions all the water must run from the equator to the poles, to get nearer the centre of attraction: which does not accord with the appearance of the planet; and it is satisfactory to find that later observers agree that Mars has no sensible ellipticity.

5. **Asteroids.**—After Mars there is a great gap among the old planets, as you see from the distance which I mentioned just now of the nearest of them, and there is no such gap in the distances beyond. But on the first day of this century, 1 January, 1801, began the discovery of a batch of little planets, or fragments of one which had been blown to pieces, which have now reached more than 100 in number, after standing for a good many years at 4. They are called the asteroids, which means things like stars, but should rather have been called *planetoids*. The first four and rather the largest are named Vesta, Ceres, Pallas, and Juno. The

others have almost exhausted the names of all the heathen goddesses, and they are now generally indicated by mere numbers enclosed in a circle, as ⑥. They are all very small, the first two being under 230 miles in diameter, and some too small to measure. As the largest is $\frac{1}{9200}$ as large as Mercury, and the moon would make 706 of it, it is evident that all of them together would only make a very insignificant planet. They lie scattered about between the distances of 240 and 300 million miles from the sun, and their periods accordingly vary from $3\frac{1}{4}$ to $5\frac{1}{2}$ years, by one of Kepler's laws, which I will explain afterward. Some of them have orbits much more inclined to the ecliptic than any of the regular planets, which further helps the supposition that they are bits of a planet blown to pieces; and so does the fact that some of them appear not to be round. Gravity there must be so small that a man could jump many times his own height.

6. After these little asteroids comes Jupiter ♃, a planet of a very different order from any we have seen yet, 1246 times bigger than the earth, and about $\frac{1}{1000}$ as large as the sun. But he, like the sun, is made of something not much heavier than water, or ·242 of the earth's density; for he is only 302 times as heavy as the earth, and $\frac{1}{1048}$ as heavy as the sun. The sun's diameter is nearly ten times Jupiter's, and Jupiter's eleven times the earth's. Notwithstanding his great size he turns round on his axis in five minutes under ten hours, and consequently the centrifugal force is so great that his equatorial diameter exceeds his polar axis by one 16th. If his density were uniformly what

it is on the average, only a quarter of the earth's, his ellipticity would be about one 9th.   Or if it increased inward at the same rate as the earth's (whatever that may be) his ellipticity would be one 13th, according to what I said respecting Mars.   As it is only a 17th, Jupiter's density must increase inward even more than the earth's, and his outside is much lighter than water. That being his ellipticity, his bulk is a 17th less than a sphere of his equatorial diameter of 86,936 miles, and about an eighth more than a sphere of his polar diameter (pp. 24, 156).

Jupiter's mean distance from the sun is 475½ million miles, or 5·2028 times the earth's, and his periodic time or year nearly twelve of ours, or 4332·6 days. Consequently his rate of travelling through space is 27,000 miles an hour, and the sun's apparent diameter there is only 6' 6''; and his average heat a 38th of ours.   As his density is nearly the same as the sun's, the force of gravity on his surface bears nearly the same proportion to that on the sun's surface as their diameters do (see p. 65), and is 2·7 times as much as on the earth; or a man on Jupiter would feel nearly three times as heavy as on the earth.

Jupiter stands nearly upright in his orbit; that is, his equator is only inclined 3° 4' to it, and only 1° 19' to the ecliptic.   The former is the inclination which affects the seasons, except so far as they depend on the changes of distance from the sun; and it is too little to make any sensible difference between summer and winter.   But as the eccentricity of his orbit is ·048, he is nearly a tenth of his mean distance, or 46 million miles, nearer the sun at perihelion than at aphelion; and as the heat varies inversely as the square of the distance, he gets a fifth more heat at perihelion than at aphelion; which is the same proportionate difference as if our summer was 110° hotter than winter: remembering that we must reckon from the 'absolute zero' of probably 522° below freezing, as at p. 43. The change of seasons in Mars, with an eccentricity of ·093, must be still greater; its perihelion heat exceeding that of aphelion as 1·37 to 1, or more than the difference of freezing and boiling water; independently of his axis being more oblique than ours. And in Mercury it is 2·23 times hotter at perihelion than aphelion; or they have a change nearly = the difference between frozen quicksilver and melted lead every six weeks.*

Jupiter not only is but looks considerably larger than any other planet, except Venus when she is nearest to the earth, his diameter varying from 30″ to 46″, or about one fiftieth of the sun's and moon's. His disc is seen in telescopes to have some dark bands or belts round it, which are always parallel to the equator, and are supposed to be clouds carried round with him; and sometimes darker spots are seen, which may be openings in the clouds. But Jupiter has a far more important characteristic than these, in his four moons or satellites, of which I will say more after we have gone through the planets themselves.

7. Saturn ♄ is not much smaller than Jupiter, but less than a third of his weight, being made of something

---

* In all these estimates the influence of atmospheric vapors is disregarded.

only ·124 as dense as the earth, and only two-thirds as heavy as water. His diameter is 73,590 miles, or rather more than nine times the earth's; but though he is 730 times as large as the earth, he is only 90 times as heavy, and the sun is 3502 times his weight. He is even more spheroidal than Jupiter, his polar diameter being one eleventh less than his equatorial, though his density is much less and his velocity of rotation rather less; for he turns in 10½ hours. His equatorial parts are supposed to be drawn out further by the attraction of the Ring, which I will mention presently. His apparent diameter is generally 18″, as he is too far off for the diameter to be much affected by the earth being in one part of her orbit or another. His mean distance from the sun is 872 million miles, or 9·539 times the earth's, and his year is 10,759 days, or 29½ of ours (the same number as the days of a revolution of the moon); consequently he moves through space about 20,000 miles an hour, and the sun appears only 3′ 20″ wide there. His light and heat are only one 90th as great as the earth's, and gravity on his surface is very little more than on the surface of the earth. Saturn stands very differently from Jupiter, with his equator inclined 26° 50′ to his orbit and 28° 10′ to the ecliptic, to which his orbit is inclined 2° 29′; and its eccentricity is ·056.

Long Inequality of Jupiter and Saturn.—I have now to describe a disturbance of these two planets by each other, of which it is difficult to know how much to attempt. When I tell you that Mr. Airy says ('Gravitation,' p. 150) that 'the calculations necessary to discover the effect of it are probably the most complicated that physical science has ever required,' you will see that nothing beyond a very general account can be expected here. If you wish to follow the subject as far as it can be carried without mathematics, you will find a longer explanation of it in the book aforesaid, and later ones in Sir J. Herschel's Astronomy, and in Mr. Proctor's book on Saturn; * but you must not expect to find any complete explanation of such a subject easy.

The mean angular velocities of Jupiter and Saturn, being inversely as their periods, are nearly in the proportion 'of 5 to 2, or more exactly, 72 to 29. It follows that every conjunction falls 242° 42′ beyond the last; or as if they fell on the angles of an equilateral triangle which itself revolves, the same way as the planets, at the rate of 2° 42′ in their synodical period of 19·86 years. If the angles are marked A B C in the direction of motion, the conjunctions fall in the order A C B A, coming round again to the same place at the third conjunction, except that it has then moved forward 8° 6′. The real motion is more irregular than this, because the velocity of the planets varies in different parts of their elliptic orbits. But if that were all, the variations would compensate each other, and all that would happen would be this:—as Jupiter ap-

---

* I think it may be a public benefit to inform those who have to employ engravers that their habit of making the letters in diagrams and architectural drawings as small as they can has made some of the plates in Mr. Proctor's valuable book totally useless to any common eyes, and even short-sighted eyes can hardly make them out. The present fashion of pale and thin printing is equally stupid.

proaches every conjunction he would be pulled forward a little by Saturn, who would be himself pulled back, and the contrary as they leave conjunction ; and the effects of the disturbances would not accumulate.

But besides the unequal velocities, the eccentricity of the orbits makes the distances also not quite the same before and after conjunction; and therefore the disturbances at any series of conjunctions may or may not compensate each other. Moreover the orbits are not in the same plane, and that causes another variation of their mutual attraction ' resolved ' in the plane of either of them ; also the nodes where they cross are continually shifting, as our equinoxes do in precession. Mr. Airy says it is impossible to do more than state that the mathematical result of all this is that for about 460 years the major axis of one planet's orbit is getting lengthened, and therefore its period (which always depends on the major axis only) is lengthened, and that of the other shortened; and then for another 460 years the effect is reversed. Though the alteration is exceedingly small in one synodical period, yet by the end of the 460 years it accumulates, like the lunar acceleration (p. 146), into something considerable, viz., an alteration of Saturn's longitude by 48′ and of the heavier Jupiter's by 21′. The eccentricities of the two orbits are also disturbed, and the perihelion of each is made to advance and recede alternately for 425 years.

Similar coincidences of periods exist, as I said, between Venus and the earth, with 5 points of conjunction in the circle, instead of 3 ; and others not so exact between other pairs of planets. But none of them produce nearly so large a disturbance as this of Jupiter and Saturn, which accumulates for such a long time, and is therefore rightly called either the *great* or the *long* inequality. You will see hereafter that a similar, and relatively a greater, disturbance exists among three of Jupiter's satellites, whose conjunctions recur at only one place in the orbit of each pair.

Saturn also has eight moons, which I postpone like Jupiter's for the present. His far more important characteristic is the Ring, which revolves round him like a continuous circle of satellites, as indeed it is probably ; and therefore that also may better be deferred till we have gone through the planets themselves.

8. Uranus.—These are all the planets that were known until the year 1781, when a new one was discovered by the late Sir W. Herschel, who was once an organist at Doncaster, and afterward the greatest astronomer of his time, and by a piece of rare good fortune the father of another not inferior to himself, and we may now add, the grandfather of a third astronomer. He, like Newton and Galileo, invented and made telescopes of his own, larger and more powerful than had been ever made before; which have been since copied on a still larger scale by Lord Rosse.

With one of his smaller telescopes Herschel found a new planet further off than Saturn, and too far to be seen without a strong telescope, though it afterward appeared that it had been seen before, but not discovered to be a planet. At first it was called the Georgian star, after the king in whose time and neighbor-

8*

hood (viz., at Slough near Windsor) it was discovered,
and by whom Herschel had been liberally assisted.
But the public preferred to name it after its discoverer,
until at last they both gave way to another heathen
god Uranus, the father of Saturn and grandfather of
Jupiter, and whose name in Greek means the heaven
itself, beyond which it was supposed there was nothing
further to be found.

   This planet Uranus ♅ or ♅ is 1753 million miles
from the sun, or 19·1824 times the earth's distance,
and takes 84 years and 7 days to go round him, at the
rate of 14,500 miles an hour.   He also ranks as a large
planet; for his diameter is nearly half Saturn's, or
33,836 miles, and he is 70 times as large as the earth,
but only 15½ times as heavy, his density being ·22 or
rather less than Jupiter's.   He is 20,470 times lighter
than the sun.   His orbit is the nearest to the ecliptic
of them all, being only inclined 46′ 29″; and its eccentricity ·046.   He has four, and some think seven or
eight moons, which behave differently from all others
in the solar system, having their orbits as nearly perpendicular to the ecliptic as 79°, and nearly circular
and moving the opposite way to all the other moons
and planets.   His apparent diameter is only 4″; and
consequently his inclination and time of rotation are
not yet ascertained.   But his spheroidicity is thought
to be as great as Saturn's; and it is most likely that
his equator is nearly perpendicular to the ecliptic, as
the orbits of Jupiter's and Saturn's moons nearly coincide with the equator of their planet.   This also accounts for his appearing sometimes spherical and some-

times spheroidal, according as his pole or equator is presented to us. The light and heat there must be 330 times less than here; gravity one-seventh less; and the apparent diameter of the sun 1' 42".

9. **Neptune.**—Still the solar system was not exhausted, as it was supposed to be when Uranus was so named. All these planets were discovered by being seen, and seen to move; but one more was proved to exist without being seen, and afterward found by looking for it where the discoverers said it would be found. The history of this discovery is so remarkable, and at the same time is given so imperfectly, and sometimes so unjustly, in larger books than this, that I shall relate it more fully than would otherwise be necessary, taking my account of it from the most authentic source, viz., the correspondence and statement of the Astronomer Royal, in vol. xvi. of the Astronomical Society's Memoirs : which very few writers of the English language, on either side of the Atlantic, and of course no French ones, have apparently taken the trouble to read before publishing their own versions of the transaction.

Just forty years after the discovery of Uranus, astronomers began to complain that he did not appear in his proper place, as calculated from the earlier observations, by which his orbit was supposed to be as well ascertained as that of any other planet. And some people went so far as to doubt whether Newton's law of attraction might not be subject to variation at so great a distance from the sun. You may be curious to know what sort of error it was in the motion of Uranus that caused so much uneasiness. By the year 1830 his

longitude, or distance from the equinoctial point ♈,
had got wrong by 30″, which would make him appear
wrong in his time of crossing the meridian of any observatory by two seconds of time; and he had got 2′ or
8 seconds wrong in 1845. This does not seem much
to be disturbed about in a planet 1753 million miles off,
or to make the laws of the universe suspected by some
astronomers, and the existence of an unseen disturber
of the peace of Uranus by others. But so it was; and
it may give you some idea of the accuracy now expected
in astronomy.

' The first person who appears to have openly suggested
the idea of Uranus being disturbed by a more distant
planet was the Rev. T. J. Hussey, of Hayes, who wrote
to the Astronomer Royal to that effect in November,
1834; and he said that two foreign astronomers, A.
Bouvard and Hansen, agreed with him. But Mr. Airy
answered that he 'did not think the irregularity of
Uranus was in such a state as to give the smallest hope
of making out the nature of any external action on the
planet,'—if there was any, which he doubted; and
preferred supposing that the earlier observations had
been wrong. In 1837 E. Bouvard, the nephew of A.
Bouvard, again wrote to Mr. Airy suggesting the same
cause; who again answered (in substance) that he did
not believe it, and added—'if it be the effect of any
unseen body, it will be nearly impossible ever to find
out its place.' By 1842 Bessel and other eminent astronomers seem to have avowed the same opinion as
Dr. Hussey, but without convincing the English Astronomer Royal, who still appears to have had no

solution of his own, except his guess at the inaccuracy of observations, although the attention of astronomers had now been directed to it for twenty years, and the error had been getting worse.

But suggesting that there must be a planet somewhere was a very different thing from setting to work to calculate whereabouts in all space it must be, with a strong presumption only, from the distances of the others, that it would be nearly twice as far from the sun as Uranus, and near the ecliptic like the rest. In 1844 Professor Challis, the head of the Cambridge Observatory, wrote to ask the Astronomer Royal for some Greenwich tables of Uranus for a 'young friend of his, Mr. J. C. Adams, of St. John's College, who was at work upon the theory of Uranus.' Mr. Airy of course sent them; and in September, 1845, Professor Challis wrote again to say that 'Mr. Adams had completed his calculations of the perturbations of Uranus by a supposed ulterior planet.' In October, 1845, Mr. Adams himself left at the Greenwich Observatory what Mr. Airy justly called afterward, 'the important paper,' giving the result of his calculations and *the place where the new planet would probably be found.*

Still Mr. Airy could not believe that a young man, who had only taken his degree (of senior wrangler) the year before had actually found the place of a planet which he believed not to exist at all, and to be 'nearly impossible to find' if it did. So instead of encouraging Mr. Adams, or taking steps to get the planet looked for by the best telescopes of various observatories, he sent him a question which he called an *experimentum crucis*, or what is popularly called a posing question. He afterward expressed his 'deep regret,' not at having done so, but that Mr. Adams did not answer it; and the world has not been informed what delayed the answer: whether it required time to answer it as completely as Mr. Adams wished; or whether the maker of what has been called the greatest astronomical discovery since Newton, felt that his announcement of it might have been better received by the public representative of English astronomical science.

But while Mr. Airy was waiting to believe in the discovery, an eminent foreign astronomer stepped into the field and confirmed it; for in June, 1846, M. Le Verrier gave to the French Academy his own independent calculations for a new planet, which nearly agreed in their result with those which Mr. Adams had given in 1845. As soon as they came here, Mr. Airy's doubts vanished, although M. Le Verrier had no more answered his question than Mr. Adams; and then he confessed (what indeed his question showed before) that he had doubted the accuracy of Mr. Adams's investigations until he received M. Le Verrier's confirmation of them; which does not mean that he considered the calculations wrong in any definite way, but simply that he doubted any man's ability to make them.

Then he did set to work to get the planet looked for in the place indicated; and Professor Challis at once undertook the search with the great Cambridge tele scope, and soon found what turned out to be the planet, and he noted it as 'appearing to have a disc,' which only planets have. But unfortunately he had no star map to compare his observations with; and also delayed comparing his own successive observations with each other, until Dr. Galle of Berlin had not only found the planet, but found that he had found it, on 23 September, 1846. Not that the finding of a planet where you are rightly told to look for it is any great feat, or at all parallel to Herschel's finding of an unsuspected planet. Only one more sentence need to be quoted from Mr. Airy's certainly candid statement to complete the story, and to show how Mr. Adams and this country lost the *undivided* credit of this great discovery, which he unquestionably first made and first disclosed. Mr. Airy says, 'I consider it quite within probability, that a publication of the elements (of the planet's orbit) obtained in October, 1845 (and given to him by Mr. Adams then), might have led to the (telescopic) discovery of the planet in November, 1845,' seven months before M. Le Verrier disclosed his calculations.

The name of Neptune ♆ was soon given to the new planet, on the same principle as the others; only they were obliged to go back to the brother of Jupiter, as he had no more ancestors in the Pagan mythology. Neptune was afterward found to be 2745 million miles from the sun, or 30·0363 times the earth's distance, and to have a period of 60,127 days, or 164⅔ years; and 41 of his revolutions take the same time as 81 of Uranus's. He only goes 10,500 miles an hour, or just $\frac{1}{10}$ as fast as Mercury; and the sun appears there only 1′ in diameter, or no larger than Venus sometimes does to us. Neptune's diameter is believed to be a little larger than Uranus's, viz., 38,133 miles, and his bulk consequently

115 times the earth's. The present estimate is that his weight is seventeen times the earth's, or that the sun is about 18,780 times his weight. His orbit is almost as circular as Venus's, and 1° 46' inclined to the ecliptic. Nothing is known of his time of rotation, or the inclination of his axis ; but though his apparent diameter is only 2', he is said to be visibly spheroidal, which implies a quick rotation. The light and heat there cannot be above a thousandth of what they are here, and gravity about a quarter less than on the earth's surface.

One satellite of Neptune has been already discovered by Mr. Lassell, revolving in 5d. 21h., and going more distinctly retrograde, or opposite to the usual direction, than the satellites of Uranus, because its orbit is only 29° inclined to the orbit of the. planet. It will be curious to ascertain whether his rotation is retrograde also. It also turns out that the planet had been occasionally seen before as a very faint star ; but as it was never seen twice in the same place, the observations had been hastily treated as mistakes ; a warning to all men never to disregard any new fact, until they are quite sure that it is not one, or is really unimportant : beside that other warning to men in high places, which the history of the discovery clearly enough proclaims.

10. There is yet one more planet suspected to exist, a very small one, only 14 million miles from the sun, and going round him in 19¾ days, in an orbit 13° inclined to the ecliptic. It was discovered (if at all) by a French physician named Lescarbault in 1859, and M. Le Verrier seems to consider its existence not improbable. The name of Vulcan was assigned to it.

But its existence becomes more improbable every year
that passes without its being seen again.

. Bode's Law of Distances.—I have several times spoken
of the distances at which the new planets of this century were expected to be found, from the proportionate
distances of the older ones; and you will see in running over them that there is a rough approximation to
a successive doubling of the distances of all beyond
the first. If you divide them all by 9 millions, to get
rid of a great number of figures, they will come in
pretty nearly the following proportions: Mercury 4,
Venus 7, Earth 10, Mars 16, Ceres and Pallas and
some of the other asteroids 28, Jupiter 52, Saturn 100,
Uranus 196, Neptune 305. Or Earth is twice as far
beyond Mercury as Venus is, Mars 4 times as far, Asteroids 8, Jupiter 16, Saturn 32, Uranus 64; but Neptune is not 128 times as far, but only 102; that is, he
is 700 million miles too near according to this rule,
which is called Bode's law, from its discoverer. And
so the only thing like a rule which there was to guess
Neptune's distance by, turned out wrong. Nevertheless it had been useful as a first approximation; for it
is a common practice in astronomy, when several things
are unknown together, to assume a probable value for
one of them, and then correct them backward after
some of them are found. Moreover the direction in
which Neptune was to be looked for was rightly calculated, which was the main point. The error in distance affected his apparent place very little, as he was
then nearly in a line with the sun and earth, and his
distance is 30 times the earth's.

You must understand that this rule of Bode's is only *empirical*, which means founded upon trial or experience—more or less extensive as the case may be—; and has nothing to do with that great law of universal attraction or gravitation which keeps the planets in their orbits, moving with a certain velocity, and makes their years bear a certain proportion to their distances, and enables their weights to be calculated from their disturbances of each other and of their moons, and an unknown planet to be found by calculation, and raises the tides, and prevents the whole earth and the sea, and everything that is in them, from flying to pieces like the wringings of a mop, and running off in straight lines into infinite space for ever.

Retrogradation of Planets.—Although the planets seen from the sun would appear to go round him, as they do, with very little variation in their pace, they appear from the earth to go very differently and irregularly in speed, and sometimes actually to go backward.   The angular velocities round the sun are inversely as the periods : so the angular velocity of the earth is always nearly 165 times Neptune's, and 84 times Uranus's, and 29½ times Saturn's, and 12 times Jupiter's, and twice Mars's, and two-thirds of Venus's, and a quarter of Mercury's.   But the angular velocities round the earth follow no such rule, and indeed no rule at all that can be expressed without a great deal of calculation : all we can do is to explain why both the interior and exterior planets sometimes appear to us to retrograde in their motion.

First take an exterior planet, Jupiter, and let him

and the sun be on opposite sides of the earth, which is called ' Jupiter in opposition ; ' and remember we have nothing to do with the daily rotation of the earth for the present. Then how does Jupiter appear to move ? He is really going, like the earth, from west to east (looking south) or from right to left, or opposite to the way the hands of a clock go ; but he goes more slowly than the earth—so slowly that he would fall behind the line of equinoxes if he had started on that line ; which you may consider to run through the earth and to be carried along with it through the heavens, moving over the infinite plane of the ecliptic, and always keeping parallel to itself, but for the trifling change of the ' precession.' Consequently Jupiter will then appear to go backward, or in the opposite direction to the sun ; not merely slower than the sun, but so much slower as to appear *retrograde in longitude :* which is reckoned forward up to 360° from the equinoctial point ♈ for celestial bodies, along the ecliptic, not along the equator from an arbitrary meridian like Greenwich, as terrestrial longitude is. So a slow-sailing ship appears to go backward from a faster one which is passing it. Jupiter and all the other exterior planets accordingly are ' retrograde ' for a few months nearly every year while the earth is passing them.

The two inferior or interior planets do the same, when they are between the earth and sun, which the exteriors of course never are. It will be easier to understand if you disregard the curvature of the orbits, and consider Venus and the earth moving in parallel lines eastward past the sun, as they do for a short time when Venus is between the earth and sun. The earth moves eastward nearly 66,000 miles an hour; and therefore the sun appears to move at that rate westward, by the stars, though not by what we call 'east and west' on the earth; or measuring by longitude in the ecliptic, he advances 2½′, the angle corresponding to an arc of 66 on a radius of 91,404. But Venus moves 77,000 miles an hour eastward, and therefore appears to move 11,000 the opposite way to the sun, or retrograde; and as she is then 25,292,000 miles from us, she apparently recedes in longitude through the angle corresponding to 11 divided by 25,292, or 1½′ an hour, for some days every time she passes between the sun and earth, or at every synodical period (see p.108 and table at the end). On the other hand all the planets appear to go faster than they do on the opposite side of the sun; and at some intermediate places they cannot be seen to move at all, or appear stationary.

SATURN'S RING AND SATELLITES.

Saturn's Ring, treating it first as single, is very thin and flat, like a ring stamped out of a card, having an outside diameter rather more than twice as wide as the planet, or 166,000 miles.* Its inside diameter is 109,-130, or only 17,660 miles from Saturn's equator, which is 73,590 miles across; and therefore the breadth of the ring itself is 28,335 miles. It is so thin that it is difficult to say what its thickness is; but it is now con-

---

* I reduce all the measures of the rings given in the latest books, in the same proportion as the diameter of Saturn, to suit the new sun's parallax. Different books give rather different measures, and nothing turns upon their accuracy.

sidered to be not more than 100 miles. And the strangest thing about it perhaps is that it seems continually to get thinner and wider, and sometimes breaks out into fresh divisions apparently, besides the long recognized and well marked division into two; of which the outer ring is 9450 miles wide, the inner 17,-300, and the space between them 1690.

If Saturn stood like Jupiter, with his equator (and ring) nearly in the plane of the ecliptic, we should have known next to nothing about the ring; for we should have seen nothing but two bright lines like handles, each of them about as long as Saturn's radius. Sometimes the satellites appear like beads threaded on the thin line of the ring. Sir W. Herschel found, by observing the motion of certain lumps or inequalities on the edge of the ring, that that part of it at any rate revolves round Saturn in 10h. 32m. 15s. or 3 minutes longer than his own time of rotation. But as Saturn's equator and ring are now* inclined 28° 10′ to the ecliptic, we sometimes see it in oblique perspective, and therefore as an ellipse, getting a very good sight of the whole width of the rings and the spaces between them each and the body of Saturn. If you want to realize this, get somebody to hold a globe in the ordinary wooden frame at some distance from you; when your eye is in the plane of the wooden horizon you can

---

* The inclinations of all the planets' equators to the ecliptic depend (1) on their inclinations to their own orbits, (2) on the inclination of those to the ecliptic, and (3) on the position of their nodes or equinoctial points; but all the nodes revolve so slowly, and the inclinations of the orbits are so small, that I do not complicate the descriptions with any further notice of these changes; but I have given the present longitudes of perihelion in the table at the end.

only see its edge ; but when it is inclined a great deal, you can see the whole of it except the part behind the globe, and you see also the space between them. The minor axis of the perspective ellipse of the ring when most elevated is ·47 of the major axis.

But the sun also must be elevated above the plane of the ring, and on the same side as the earth, though not necessarily as much, to illuminate the side of the ring facing the earth, to enable us to see it. You will find in Mr. Proctor's book pictures of Saturn and the rings in all possible phases, with a variety of interesting descriptions of them which it would be out of place to give here. If Saturn has inhabitants they enjoy an eclipse of the sun by the ring for about 15 years, or half Saturn's period, over a considerable width of his surface.*

The mass of the ring has been calculated as the 118th part of that of Saturn, from its effect in disturbing some of the satellites. And that agrees pretty well with the estimate of 100 miles of thickness, on the assumption that its average density is the same as Saturn's own. There have been various speculations as to its composition. Laplace proved that it could not be in one solid flat piece, but must be at least divided into two ; and the same kind of reasoning has made it necessary to carry the division still further. For the outer and inner parts of a ring of anything like that width require different velocities and periods to preserve their equilibrium, according to well-known laws of motion which will be explained in the next chapter. If the inner part only went as fast as the outer, its cen-

---

* Appendix, Note XIX.

trifugal force would not be enough to keep it from being dragged into Saturn by attraction. For the ring itself is far too thin and weak to be able to hold itself together by its own cohesion or internal attraction, against such a force as that tending to pull it in pieces. Moreover the outer edge revolves in the proper time for a satellite at that distance; but too slow for one at the inside, or even at the middle of the breadth of the ring.

But a far more serious objection to the rings being rigid at all has been made by Prof. Peirce, of Harvard University.* Mr. J. C. Maxwell, of Trin. Coll., Cambridge, received the Adams Prize Essay of 1857 for the discussion of a special case which was embraced in Prof. Peirce's previously published and exhaustive paper, viz., that, in order to preserve its equilibrium in revolution, each ring must be so uneven in density that its centre of gravity may be more than 9 times further from the light side than the heavy one, if it is rigid; which is completely at variance with observation—unless there is that enormous and improbable latent difference in the opposite sides. Moreover the rings would then be much nearer the planet on one side than the other, which they are not, though they are a little. You may ask how a ring can be stronger for being *limp* than rigid, provided it is of the proper shape, whatever that may be. The answer is that there is no proper rigid shape, not even that defined by Mr. Maxwell: which certainly does not exist; and·if it did, it could not maintain itself against disturbances. Consequently a set of rigid narrow rings is no less impossible than a single wide one.

---

* Appendix, Note XX.

The idea has been therefore entertained that the rings might be fluid, so that all their parts could move along each other as they pleased.   But to this also it is objected that the constant changes of motion would cause waves, which would break the rings in pieces, for the effect of waves lasts long after the force has passed away which raised them.   Moreover the appearance of the rings contradicts that theory.   For beside the one, or we must now say two, very evident divisions, there are indications of more, and especially of an inner ring, discovered by Prof. Bond of Harvard University, darker than the others, but semi-transparent, as if composed of bodies near together but not too close to see through.   And in fact that is now taken by astronomers to be the constitution of all the rings of Saturn. Each ring, of only one satellite in width everywhere, whether distinctly or indistinctly separated from the rest, is thought to be a vast number of satellites moving in one orbit, and therefore in the same time; as everybody has read lately there is reason to believe that several rings of meteors travel round the sun in orbits which cross ours at certain days in the year.

Another fact quite inconsistent with rigidity, is that the rings as a whole have been getting thinner, and about one-sixth wider, the inner parts coming nearer to the planet, but the outer not altering, in the 200 years since Huyghens first discovered and measured it; and more rapidly in the 80 years since Sir W. Herschel's time than in the 120 years before.   Galileo saw the ring but could not make it out: he did distinguish Jupiter's satellites with his second telescope in 1610.

Saturn's Moons are far less interesting than his rings. It is enough to say here that the eighth and last in the order of discovery, but the seventh in distance and a very small one, was only found in 1848, and oddly enough, by Mr. Lassell in England and Mr. Bond in America on the same night. The sixth is much the largest, and so it has been called Titan in Sir J. Herschel's re-naming of them all by name sinstead of numbers—Mimas, Enceladus, Tethys, Dione, Rhea, Titan, Hyperion, Iapetus. But though some of these were the 'giants' of the ancient world who fought with Jupiter, they are very small satellites of Saturn now, some too small to be seen except with the best telescopes. Iapetus is one of the larger ones, and much further off than the rest, being 32 of Saturn's diameters distant from his centre. Our moon's distance is 30 times the earth's diameter; but then Saturn's diameter is nine times ours. Their distances and periods, which is all that seems accurately known of them, are expressed shortly in the following table. The first five follow Bode's law pretty well, but not the others.

SATURN'S MOONS.	DISTANCE FROM SATURN.		Period.	Discovered.
	In miles.	In radii of ♄		
			d. h. m.	
1 Mimas..........	123,633	3·36	0 22 37	} 1789
2 Enceladus......	158,590	4·31	1 8 53	
3 Tethys........	196,480	5·34	1 21 18	} 1684
4 Dione..........	251,680	6·84	2 17 ·41	
5 Rhea..........	352,390	9·55	4 12 25	1672
6 Titan	814,840	22·145	15 22 41	1655
7 Hyperion	985,380	26·78	21 7 8	1848
8 Iapetus........	2,367,600	64·36	29 7 54	1671

9

The periods of Uranus's moons run from 2·53 to
107·7 days, and their distances from 128,800 to 1,570,-
000 miles; the last three distances nearly doubling
each other, and the first five differing by about 50,000
miles each.   But two at least of them are doubtful.

Jupiter's Four Satellites are of far more importance
than Saturn's eight.   They are all of a substantial size :
the first, second, and fourth about as large as our moon,
but a great deal lighter, having one-third, two-thirds,
and half the moon's density; and the third having a
diameter and weight about half as much again as the
moon, and three times her bulk.   That satellite also
has two-thirds of the density of the moon, or two-fifths
of the earth's density, and half as much again as Jupiter's own.   These variations of density among the
planets and their satellites are remarkable, following
no apparent law whatever.

Their distances from the centre of Jupiter are respectively about 6, 9½, 15¾, and 27 times his radius; which
do follow Bode's law very well, the differences being
successively about 3, 6, 12.   Consequently the first appears to people on his surface (if there are any) about
as large as our moon does to us, the second and third
about half as wide, and the fourth a quarter as wide,
or one-sixteenth as large.   But the largest of Jupiter's
moons has only the 11,300th of his mass, while our
moon is nearly one-eightieth of the earth.   Their elements are more fully given in the table at the end of
the book.

They all move so nearly in the plane of Jupiter's
orbit (which also nearly coincides both with his equa-

tor and our ecliptic) that they are eclipsed every time they pass, except that the fourth escapes sometimes. And they, like our moon, always show the same face to their planet. The largest of them looks rather less than Neptune, and the apparent diameter of the smallest, at our mean distance from them, is under $1''$.

The period of the first is $42\frac{1}{2}$ hours, of the second $85\frac{1}{4}$, of the third $172\frac{3}{4}$, and of the fourth 16d. $16\frac{1}{2}$h.; so that the periods of the first three are successively a very little more than double of each other. And from that coincidence of periods several remarkable consequences follow. Whenever the second and third are in conjunction, the first is exactly opposite to them; and the place for conjunction of the first and second will be opposite to that for the second and third. Consequently they cannot all be either eclipsed from the sun, or hidden from us by Jupiter, or made invisible by being in front of his disc with the sun shining on both it and them. But two may be eclipsed or hidden while the other is so made invisible; and as the fourth may be anywhere, that may also be invisible at the same time in any of those three ways. One of these very rare *non-phenomena* took place on 21 August, 1867, from 10.4 to 10.49 P. M.

Another consequence is, that all the three orbits are mutually affected by the disturbances, or strongest mutual attractions, constantly recurring at the same place for the first and second, and at the opposite place for the second and third (which aggravates the effect upon the second). The effect is to make the first and second orbits ellipses with the apses always on the line of conjunctions; and a kind of secondary major axis or pair of apses of the third orbit also: only that happens to have a larger independent eccentricity of its own in another direction. But the *perijove* of the first and the *apojove* of the second are at their place of conjunction; and the second's perijove and the third's apojove are at *their* conjunction; or the three orbits are pushed away from each other at the two places of conjunction.

But the periods are not *exactly* double of each other; and that makes the line of conjunctions and of apses revolve slowly backward as each conjunction comes a little before the place of the last; and it takes 486½ days for them to work round, which makes 68 periods of the third satellite, and 137 of the second, and 275 of the first. This is something like the 'great inequality' of Jupiter and Saturn (p. 174): indeed this is relatively greater, for it takes many more periods of the disturbed bodies to work round. Moreover here *every* conjunction falls near the same place, but only every *third* conjunction of Jupiter and Saturn, and the two intermediate ones rather counteract than aggravate the effect. This subject is worked out at greater length in Mr. Airy's Gravitation than we can afford to it here; and it is not an easy one.

The satellites of Jupiter, and Saturn too, suffer another disturbance from the great oblateness of those planets, and from Saturn's ring. Let us see how the attraction of a sphere on a satellite in the plane of its equator is altered by shaving pieces off the poles and laying them round the equator, so as to make it into an oblate spheroid. The piece laid on the side nearest to the moon will attract more than it did at the pole, because it is both brought nearer and also into the direct line of attraction, or the line of centres. The piece laid on the far side loses force by being put further off, though not quite so much as the other gained by being put nearer (see p. 134) and gains by being brought into the line of centres. The attraction of the pieces which are moved from the poles to the equidistant sides of the equator is not sensibly altered. Therefore on the whole the equatorial attraction of the spheroid is greater than of the sphere, on a moon at the same distance from the planet's centre, though it is less on his own equator, being further off (p. 37).

The distance of the first moon being 6 times Jupiter's radius, his attraction on it would be thus increased by nearly a 24th of his ellipticity, or a 408th, and would shorten its period an 816th, if his outside were as dense as his inside : which it is not (p. 172), and therefore the effect is less. (This calculation does not hold for attraction on the surface, nor at small distances). The effect is less on the fourth moon, which is 27 radii off, in the proportion of 6^2 to 27^2, or not one 20th as much. Therefore also each moon is less accelerated at apojove than perijove ; and that *comparative* loss of attraction at apojove makes their apses advance, as the greater loss of earth's attraction at apogee than perigee makes the apses of our moon advance (see p. 152).

The spheroidicity of the earth produces similar effects on the moon, but too small to be appreciable in her period, because the earth is much less oblate than

Jupiter, and our moon's distance is 60 times the earth's radius. But on the other hand, she goes much higher above the equator than his moons do, viz. 28½° ; and consequently she suffers another disturbance in return for her disturbing our polar axis by nutation (p. 55). For the polar attraction of an oblate spheroid at a given distance falls short of that of an equal sphere twice as much as the equatorial attraction exceeds it, though the attraction *at* the pole is greater (p. 37). Therefore the earth's attraction on the moon is greatest when she is on or near the equator; and thus she is disturbed both in latitude and longitude by the earth's ellipticity: which can be calculated from the amount of these disturbances, and is found to agree with the result obtained by other means.

The eclipses of Jupiter's moons are seen on his right or left side according as the earth is right or left of the line from the sun to Jupiter, which prolonged is the axis of his shadow.   Besides the eclipses by his shadow and *occultations* behind him, the satellites may *transit* over his face, either as dark or bright spots, according to the position of the sun, and according as they pass over a dark belt of Jupiter or a bright part of his face; or they may cast their shadows as dark spots upon him while their bodies appear bright, either on his disc or beyond it.

Velocity of Light Discovered.—Some years after Galileo had discovered the satellites, their eclipses were observed always to come too soon when Jupiter was at his nearest to the earth, and too late when he was furthest off.   The extreme difference between the early and the late eclipses was 16m. 26s., a variation far too great to be tolerated or attributed to mistakes even in the early days of astronomy. Accordingly Römer, a Danish astronomer, in 1675 hit upon the solution that the light itself takes some definite time to come, and spends those 16½ minutes in coming across the whole width of the earth's orbit, or the difference between the nearest and the furthest distances of Jupiter and his moons from us. Therefore the light takes 8¼ minutes to come here from the sun, which is at half that difference of distances. This does not tell us what the velocity of light is until the sun's distance is ascertained, which we shall see how to do presently; or if the velocity of light is found by other means (see p. 86), this will tell us the sun's distance. But we have not quite done with Jupiter's moons yet.

Finding the Longitude.—The longitude of a place is simply the difference between the local clock time and the clock time of Greenwich, turned into degrees at the rate 4 min. to 1° (p. 19). The time of any place is that of a clock which is at 12 when the sun is on the meridian of the place, subject to the correction called the *equation of time,* which is the same for all places, and is given in the almanacs for every day (see p. 58). The time of noon at any place may be found from the fact that the sun is equally high above the horizon, and shadows are of equal length, at equal times before and after noon; besides other methods involving more mathematics, and by the stars as well as the sun. The beginning or end of an eclipse of a satellite of Jupiter, which happens very often, may be used as a common signal to be observed at Greenwich and the place in question, as much as if it were a rocket visible to them both, or the local time of one telegraphed to the other. Instead of the eclipses being observed at Greenwich, they are calculated for Greenwich time, and published several years beforehand in the Nautical Almanac; and so when you have observed the local time of an eclipse anywhere, you can at once see its difference from the Greenwich time. But after all this method is little used in navigation, because these eclipses cannot be observed accurately enough on board ship, and it is superseded by that of

Longitude by Lunar Distances.—The apparent distance of the moon from the sun, or any convenient star, can be measured by the seaman's instrument, called a *sextant,* first invented by Newton, in which you see one object coinciding with a second reflection of the other from two small mirrors, when one of them is turned through the proper angle, which you can then read off; and a man can hold it in his hands steadily enough.* This apparent distance has next to be turned into *true distance,* which means the apparent distance as it would be seen from the earth's centre (see p. 211), by calculations for which tables are provided : and the local time of the observation must be taken. The true distance of the moon from the sun and some suitable stars is given in the Nautical Almanac for every 3 hours of Greenwich time through the year; from which the Greenwich time corresponding to the observed distance (reduced to true) can easily be

---

* See Newton's paper in Ph. Tr. vol. 43, and *Sextant* in English Cyc.

found; and its difference from the local time of the observation is the longitude of the place.

If you have some chronometers keeping Greenwich time, no other observation is wanted but that of the local time of noon, or of any star transit whose Greenwich time is in the almanac. It is not safe to rely on one watch, and if two differ you do not know which is most right; but three give a tolerably safe average, if they are good ones; and of course the more you have the better. Sometimes a great many are used for fixing the longitude of important places, and are carried backward and forward between them and Greenwich. The parliamentary reward of £20,000 was given to John Harrison in 1767, for making the first chronometers that would find the longitude within 30 miles after a long voyage; and they have been very much improved since then. The makers send chronometers to the Royal Observatory for six months' trial, and the Admiralty buys some of the best of them every year. But the trials are too much confined to variations of temperature, far beyond what the chronometers are practically exposed to, even in Arctic expeditions: see the Rudimentary Treatise on Clocks, p. 299.

Finding the latitude is a more straightforward operation, and there are several ways of doing it. For the latitude of a place is the angular distance of the equator from the zenith, along the meridian; which is equal to the sun's zenith distance at noon, added to his *declination* or distance from the equator, or 90°—his *polar distance*, which is in the almanac for every day.

TRANSITS OF VENUS AND MERCURY, AND MEASURING
OF SUN'S DISTANCE.

Though Mars and Venus have no moon, they sometimes have the shape of one themselves, and Mercury
too ; only he is too small and too near the sun for his
shape to be often seen. But Mars never has the crescent shape, because that can only be the phase of a
body between the earth and sun, which Venus is sometimes, just like our moon, only much further off. Mars
can appear no narrower than gibbous (see p. 106), as the
moon does in the second and third quarters, or the
fortnight with the full moon in the middle. I have
already said that the sun's distance has lately been
corrected by certain observations of Mars, but the determination of it from the transits of Venus is considered to be more accurate, and is one of the most important and curious problems in astronomy, though
Venus is very seldom available for it. Mr. Airy calls
it ' a noble problem, deserving attention as well for its
ingenuity as the certainty of its conclusions ; ' but it is
now certainly concluded that a mistake was made by
one of the observers the last time it was done. He
adds that ' it is one of the most difficult subjects for a
lecture that he knows.' Nevertheless we will try what
can be done toward making it intelligible, by an explanation somewhat different from his, though of course
ending in the same result.

I had better first tell you that the *proportionate* distances of the sun, earth, and Venus, to the diameter of
the sun can be easily ascertained without waiting for

the rare phenomenon of transits. The proportion between the sun's diameter and distance is nothing more than his apparent diameter, 32′; for that one figure expresses to those who understand how to reduce degrees and minutes into numbers, that the sun's diameter is to his distance as the length of an arc of 32′ is to an arc equal in length to the radius of the circle, or 57° 18′; and the numerical value of 32′, or the sun's diameter divided by his distance, is ·00935. From that you may easily calculate that the radius (= 1) is 108·2 times the length of 32′, and therefore we are certain that the sun's mean distance is 108·2 times his diameter, whatever that may be. Next the apparent or proportionate distance of Venus¦ from the sun is found, as I have said before, by observing the angle between Venus and the sun when they are widest apart; and that being done on various occasions, when she is at different distances from the sun, her whole orbit is known; and therefore her proportionate distance from the earth and sun is known at the time when the transit is going to happen. And we may say, in round numbers, that the sun is then twice and a half as far from Venus as the earth is, though we yet know nothing of the actual length of any of their distances.

A transit of Venus over the sun is exactly the same in principle as an eclipse of the sun by a very small moon, or what is the same thing to us, a very distant one; and like our eclipses, it can only happen when Venus is at or very close to one of the nodes where her orbit crosses our ecliptic, exactly at the same time that she is crossing between the earth and sun, which is called *inferior conjunction.* (Superior conjunction is when the planet passes behind the sun, but we have nothing to do with that). Now if two men stand before a post with a wall behind it, they will see different places on the wall eclipsed or hidden by the post; and if the post is as far from the two eclipsed places as it is from the men, the two eclipses will be exactly as far apart as the two men are; if the wall is twice as far from the post, the two eclipses will be twice as far apart, and so on.

Therefore two people on the earth, as far apart as they can conveniently get for them both to see the transit of Venus from beginning to end, will see at the same time the two transit spots twice and a half as far apart in real distance on the sun as the observers are distant from each other. Suppose they are 7200 miles apart (measuring through the earth the shortest way) then the two transit spots will be 18,000 miles apart on the sun; and we have only one step more to take in order to find the diameter of the sun in miles; and that is, to get an accurate map made of the disc of the sun with the exact positions of the two spots at the same time; for then we can measure their distance on the map and see what proportion it bears to the diameter, and we know that 18,000 miles bears that same proportion to the real diameter of the sun, and the business is done.

The real difficulty is to get this sun-map made accurate enough to measure from, or to get the exact distance of the spots at the same moment, remembering that the two observers are nearly half way round the earth from each other. For that purpose the following contrivance is adopted. Instead of observing the transit at one moment only, each man observes the whole path of Venus across the sun ; or rather in reality he observes *the exact time it takes ;* for they can observe the first and last contact of the spot far more accurately than they can measure distances on the bright face of the sun; and it is not necessary that they should see anything but the beginning and the end of the transit. The places on the earth are so chosen that the paths may appear not only parallel, but at the widest distance possible apart, forming two chords across the sun, parallel to the diameter which Venus would pass along if she was exactly in the ecliptic and seen from the centre of the earth. The two paths may be on different sides of the sun's centre if Venus is exactly at a node, but they are more likely to be on the same side, in which case their difference of length is greater, and the observations more likely to give an accurate result.

For the accuracy of the map depends on this: you have a circle of known diameter to start with, because the time Venus would take to cross the middle of the sun is known from the proportion which his diameter bears to the orbit of Venus, and the time she takes to perform it. So if that time were known to be 6 hours we might draw a circle of 6 inches diameter for the sun ; and if one observer reported his transit to have lasted 5 hours, we should find the place where a chord 5 inches long will exactly fit ; and if the other transit lasted $5\frac{1}{4}$ hours, we should put in another chord $5\frac{1}{4}$ inches long, parallel to and near the former. (The real lengths

could not be exactly these, but that does not signify).
The distance between·two chords of 5 and 5¼ inches in
a circle 6 inches wide can be calculated with the utmost accuracy, and also the proportion of that distance
to the diameter, which is the proportion of the 18,000
miles to the real diameter of the sun, the thing we
wanted.

I have said nothing about the rotation of the earth
during the time the transit lasts; but of course due
allowance has to be made for that by methods known
to astronomers.   What I have told you embraces all
the principles of the calculation; which was first suggested by James Gregory in 1663 (who made the first
reflecting telescopes, different from Newton's) for both
Mercury and Venus.   But Mercury is more difficult to
observe, and it was long before it could be applied to
Venus: for her last transit had been in 1639, the year
Gregory was born, and they only come in pairs of 8
years separated by 105½ and 121½ alternately.   The
last were in June 1761 and 1769, and the next will be
in December 1874 and 1882.   No doubt provision will
be made both by England and France for having the
observations made at several pairs of distant places, in
case any of them should be clouded, with greater accuracy than appears to have been obtained in 1769,
when the celebrated Captain Cook went to Otaheite
for the purpose at the expense of George the Third;
and other people went to Lapland, where the beginning of the transit was observed in the evening of a
short night, and the end the next morning: the mistake is supposed to have been made there, and not at

Otaheite. In 1761 the observation at St. Helena was spoilt by clouds.

Transits of Mercury are much more frequent, coming at intervals of 7 and 13 years; but the same cause which makes them more frequent makes them less useful, viz., Mercury's being nearly twice as near the Sun as Venus is, and twice as far from us; which makes the breadth between the two transit paths four times as narrow as for Venus, and much more than four times less likely to give a correct measure of the sun's diameter and distance.

SUN'S PARALLAX, AND MOON'S.

Another thing which is found as part of the same operation in a transit of Venus, is the apparent diameter of the earth as seen from the sun; *half* of which angle is called the sun's *parallax;* and that is merely the earth's equatorial radius of 3963 miles divided by the sun's distance of 91,404,000, which represents an angle of 8″·943. Until lately it was considered 8″·57, but it is increased one 23d with the diminution of the sun's distance by one 24th, as stated at page 87. This parallax, rather than the sun's distance, is usually taken as the standard measure for the whole solar system, though of course they go together. The parallax of Venus, or the apparent radius of the earth seen from Venus at her nearest to us, is about five times the sun's parallax, being in the inverse proportion of their distances from the earth. All the measures of the solar system may be calculated from the parallax of Venus or Mars, or any other of the planets, as the proportionate distances of them all are known independently, and only one real distance is wanted to supply the scale for measuring them all.

The parallax of the sun or moon then is the converse of their apparent semi-diameter, being the earth's apparent semi-diameter to them respectively. The moon's mean parallax is 57′ 2″, and her mean semi-diameter 15′ 33″, and the sun's mean semi-diameter 16′.

As I have said that the moon's distance is measured independently of the sun and any planet, I ought to tell you how it is done—in the best way, for there are several, as you may see in Mr. Airy's lectures and other books. Two observers as far apart as they can be for them both to see the moon at once, will see her apparently covering two different places among the stars behind her, as a post covers different places in the wall behind it from different observers; that is, she will appear to each of them at a different distance from the same star pretty near her; which distance is measured, as usual, by the angle between the two lines of sight from the observer to the star and to the moon's centre. When all proper corrections have been made, the difference of those two angles will be the apparent distance of the two observers from each other as seen from the moon's centre. Then that differential angle must be a little increased, in proportion as their real distance falls short of the diameter of the earth; and half the angle so increased will be the apparent radius of the earth as seen from the moon, and that is the moon's parallax.

You may ask why the sun's parallax should not be

got in the same way, without calling in the aid of Venus. Because stars near the sun cannot be seen, even with a telescope, and stars far from the sun will not do, for various reasons. In sunlight all objects are unsteady, from the variation of the density of the air, as will be explained with Refraction, and as you may see in a very hot day, and still better near a fire in the open air; and the unsteadiness is magnified by telescopes, so that the sun's distance from a star, either far or near, could not be accurately measured. Other methods have been used, but they are not accurate enough for modern astronomy.

The parallax of Mars however can be got with considerable accuracy in the same way as the moon's, though he is 200 times further off. When Mars is in conjunction with the earth, or opposition to the sun (shortly called 'Mars in opposition') he is only 48 million miles off, and may be under 44, if he is in opposition and perihelion together, as the eccentricity of his orbit is ·093. (The earth's aphelion is so situated that we cannot get the benefit of the eccentricities of both orbits at once.) This happens every 8 years, and it was a little before the last opposition at perihelion, in 1862, that suspicion had begun to attach to some of the transit of Venus observations of 1769, and to the received distance of the sun. Advantage was taken of that favorable position of Mars to get his parallax by a series of observations at places as far off as possible in latitude and near the same meridians, such as Greenwich, and Melbourne in Australia, Pulkowa, and the Cape of Good Hope, where an observatory has been established for observing the southern hemisphere.

The result was that all the observations agreed in giving that parallax, and therefore distance, for Mars which makes (by proportion) the sun's distance nearly 91½ million miles: which is curiously about midway between the distance obtained by similar observations of Mars in 1672, and those of Venus a century afterward. It must be remembered that although a single observation of Mars' parallax is not equal to one of a transit of Venus, properly taken, we are not limited to one or two of Mars in a century, but a great many may be taken at each opposition. No less than 60 were got in Australia in the two or three months while Mars was conveniently situated in 1862. And it may now be said for the first time in the history of astronomy that there are both many and independent proofs of the sun's distance being what is received by astronomers; for several others beside these and the velocity of light (p. 86) have since been added, and transits of Venus are no longer of the importance that they were; though it is highly desirable that the late corrections should be tested by this, which is undoubtedly the best single method of all.

' Parallax ' is also used for other things, both in the heavens and in the earth. The parallax of the fixed stars is always referred, not to the radius of the earth, but to the radius of the earth's orbit, of which I shall have more to say hereafter. Parallax may be defined to be the apparent displacement of a body from shifting your place of observation, and you may say that it varies inversely as the distance, when the distances are great. Thus if you see a fire at some unknown distance to the north, and if it still looks north when you have gone a good way east or west, you may be sure it is a long way off, because its parallax is small. So too the long hand of a church clock does not really point to 15 or 45 minutes when it seems to do so, but a little below them, because the hand is not close to the face, and so is displaced by parallax from the position you would see it in if you were on a level with it.

All astronomical observations, except those for finding a parallax, have to be corrected for the earth's parallax, or as it is called shortly—parallax, or the distance of the observers from the earth's centre. Otherwise the observations made and recorded at one observatory would be unintelligible and useless at any other. By that correction they are 'reduced' to the earth's centre as a common point for all observations alike. We have now to consider two other corrections which have to be made before any observation can be considered complete, and in a state fit to be recorded for future use.

ABERRATION AND REFRACTION.

The stars and sun and planets are apparently displaced, or are seen in wrong places, from another cause called *aberration.* The discovery of it by Bradley in 1727, the year that Newton died, was a consequence and confirmation of the previous discovery, that light takes a definite and measurable time to come from the sun and planets; not that the planets have any light

of their own, but only reflect the sun's, as the moon
does.   We may explain aberration thus : If you are
running when the rain comes down straight without
any wind, you get wet in front and not behind, and the
rain beats against you as it would if you were standing
still and the wind blowing in your face.   And if you
carry an empty telescope tube pointed straight up, the
rain will not fall through it, but will strike against the
back inside : if you want the rain to fall through, you
must slope the tube forward, more or less according to
your velocity forward compared with that of the rain
downward.   Then for rain substitute light, and the
motion of the earth for your own running, and you
know what aberration is.

Therefore whenever we are moving directly across
the light from any star, it appears before its true place
by 20″, which $= ·00001$, or the proportion of the velocity of the earth to that of light (pp. 36, 86).   When
we are approaching a star, or receding straight from it,
there is evidently no aberration.   Therefore stars near
the ecliptic oscillate 40″ backward and forward every
year apparently, in consequence of the real motion of
the earth in its orbit.   But stars near the poles of the
ecliptic are carried round a whole circle of aberration
40″ in diameter, neglecting the insignificant ellipticity
(not eccentricity) of the earth's orbit.   All other stars
have their aberration circle foreshortened into a perspective ellipse, with a major axis of 40″ lying across
the line to the earth, and the minor axis diminished
according to the star's latitude or distance from the
ecliptic, where the ellipse sinks into the line of oscilla-

tion. But aberration always diminishes the sun's longitude by 20″; for we see him by the light which left him 8¼ minutes ago when he was 20″ behind his present place.

The November meteors, which I shall describe presently, afford a much stronger case of aberration. They go the opposite way to the earth, and at rather greater speed, in an orbit inclined only 17° to the ecliptic. Therefore they meet the earth with quite as much velocity as if it stood still and they went twice as quick or twice as far forward in a second from any given height above the ecliptic; that is, at the apparent inclination of 8° or 9° instead of 17°.

Refraction.—There is yet another correction to be applied to most telescopic observations before they can be said to have given the true place of a star, as it would appear from an earth no bigger than a point, quite still, and without an atmosphere to bend the rays out of a straight line. For this is what the atmosphere does, and it is called *refraction.* If you hold a straight stick obliquely in a trough of water, the part in the water appears to be bent upward, or further from upright, with an elbow at the surface of the water, and the trough itself looks less deep than it does when it is empty; so that if you stand where you cannot quite see the bottom of the trough empty, you will be able to see it when it is full. And conversely, if your head were under water all things outside of it would appear elevated.

The reason is, first, that you see everything in the direction of the rays as they at last reach the eye, by whatever road they have come, as things seen in a mirror appear to be behind it; and secondly, the rays of light are always bent toward the perpendicular in the denser of two mediums which they pass through obliquely. Consequently anything in the water is seen by rays in the air more oblique than those which started from it in the water, and so it appears lifted. Again, as air is denser than empty space, though very much thinner than water, the rays which come to us from the stars are bent toward the perpendicular, and so they appear higher than they are, except when they are already as high as possible, or in the zenith. And the sun and moon always do, since they are never in our zenith; and the lower they are the more they are raised by refraction. For the law of refraction is, that the distance of any point of a ray from the perpendicular to the surface in the second of the two given transparent mediums always bears a fixed proportion to what the distance of the same point would have been if the ray had not been bent aside or refracted. This is shortly expressed to those who know a little trigonometry by saying that the *sine* of the angle of incidence bears a fixed proportion to the sine of the angle of refraction. I shall have to say more about this in the chapter on telescopes.

Besides this, the density, and therefore the refraction of the air decreases upward, as you may see by taking a portable barometer up a mountain; so much that, although an atmosphere of uniform density and of the known weight of 15 lbs. on the square inch would only reach 5 miles high, it does in fact reach about

80.* Consequently the rays are not bent with a single elbow, like the image of the stick in water, but into a curve, continually getting more upright as they approach the earth, and you see the star in the direction in which the rays at last reach the eye. Moreover, the lower the star is, the more obliquely the rays enter the air, and the more they are bent. Refraction is also diminished by heat, as that expands the air and makes it thinner, and it increases with a rise of the barometer, which indicates increased density in the air, and so it can only be found and applied by tables which have been prepared from long experience. Mr. Airy calls 'refraction the bane of astronomers,' because it cannot be calculated with certainty, like aberration and parallax.

The average amount of it for objects about half way up from the horizon to the zenith is about 1′, but at the horizon it is as much as 33′, which is rather more than the apparent diameter of the sun or moon. Consequently they have really set, or have not risen, when they appear to be just above the horizon, being lifted their whole height by refraction; and the moon may be totally eclipsed with the sun apparently above the horizon, though the earth is really straight between them. From the same cause the French cliffs, and even ships on the sea near them, can be seen from the English coast in some states of the atmosphere. There is no refraction sideways, and it increases so rapidly toward the horizon, that the lower edge (or *limb*, as they call it) of the sun or moon is lifted rather more

---

* Appendix, Note VI.

than the upper, and therefore they do not appear quite round, but visibly broader than high when they are just rising or setting, and yet not exactly elliptical, but more flattened at the bottom than the top.

Their appearing larger at the horizon to the naked eye is only an optical delusion; for in fact they appear smaller, as the vertical diameter is diminished by refraction and the horizontal one is not increased. The delusion is generally attributed to our being able to compare them at the horizon with things on the earth. But I doubt if this is the proper explanation ; for it is equally the case when it is too dark to see anything but the moon itself just rising, and at sea, where there is nothing else to be seen. Most people have observed how much larger a man looks against the horizon on the top of a hill than when you are at the top and he is at the bottom. I believe the reason is that when things are on the horizon we compare their linear dimensions with the length of horizon which the eye takes in, but that in the middle of the sky or earth we compare the area they cover with the area the eye takes in. Assuming the eye to see distinctly over 30° of apparent width, the moon on the horizon covers one · 60th of that; but when it is high up it only fills the 3600th of the area of sky which the eye sees all round it. Between these two extremes some compromise is involuntarily made, and a different one by different people's eyes.

Another theory has lately been propounded, that the enlargement of the sun and moon on the horizon is due to a peculiar effect of the red rays on the eye, which then evidently preponderate, from the greater absorption of the other colors by the atmosphere. But whether right or wrong, that theory seems yet in a very undeveloped state.

From another cause the moon really measures altogether less when she is rising or setting; that is, when we are just coming into or going out of sight of her. For the parts of the earth to which the moon is just rising may be further off by half the earth's diameter than the places full in front, which have the moon on their meridian; and the earth's radius is one 60th of the moon's distance. Therefore her diameter will appear one 60th less, and her whole disc about one 30th less to the places furthest off, which can just see her, than to the places nearest. But the difference of the earth's radius is practically nothing in the distance of the sun, and therefore you cannot say that the sun appears smaller on the horizon than on the meridian, as the moon does, except from refraction.

Twilight is also caused by the air, but not in the same way as refraction. When the sun is not more than about 15° below the horizon his rays are a little *reflected* down to us from the vapors and other small particles of matter in the air. The more obliquely the sun goes down the longer he takes to get as low as 15° below the horizon, and consequently twilight lasts longer in high latitudes than within the tropics, where the sun's path may be vertical or through the zenith at noon, and is quite so in some latitude there every day.

The blueness of the sky is also due to the reflection of the violet rays by the air, from some unknown

10

cause; and is therefore greater when the sun is low, and greatest at night, when he is below the horizon altogether.

METEORS OR SHOOTING STARS.

As these bodies are of the nature of very small planets, they may properly be considered next. Early in the morning of the 14th of November, 1866, 'the *great* November star shower' came according to prediction. Every year at that time there is a shower of meteors or shooting stars; but every 33 years it has a maximum. Prof. H. A. Newton, of Yale College, first propounded what may be called a meteoric theory, which Mr. Adams has lately confirmed, as you will see presently.*

Besides the November shower there is another on August 10 every year, which I will speak of first, because its period is more simple, agreeing within a minute, from the earliest Chinese Annals (in which they are recorded nearly 1000 years ago) with the length of the sidereal year, or absolute revolution of the earth without regard to the precession of the equinoxes; and therefore these August meteors come a day later in 71 equinoctial years. It does not appear that they have a maximum of 33 years or any other time. Their effect would be produced either by a continuous ring of meteors revolving round the sun in any period and any orbit which crosses the earth's orbit at the place where it now is on the 10th of August: or by a mere cluster of meteors revolving in a similar orbit with exactly the sidereal period of the earth; for if such a

---

* Appendix, Note XXI.

cluster once meets the earth where their orbits cross, it must always do so, until the revolution of apses of the orbits throws them out of contact.

If there is a continuous ring of meteors the earth must fall in with them every time it crosses the ring. If it were not only a ring, but a plate of meteors as wide as the earth's orbit, as the zodiacal light (p. 71) is thought to be, then we should fall in with it again about February 10: which is not the case. Even a continuous ring would be crossed again in February, unless it is an ellipse differing so much from the earth's orbit that they only intersect once, and miss each other where they pass at the opposite half of each orbit, which is probably the case.

All the meteors of each shower appear to radiate from one point, or more correctly, from one small circle or ellipse. The November meteors radiate from such a point in the constellation Leo. But this is only the effect of perspective. The earth drives through, and nearly at them; for they go the opposite way to the earth, and cross the orbit so obliquely, that we practically meet them, and with more than twice the earth's velocity. If a quantity of rockets were shot toward you from the far end of a street, they would all appear to radiate outward from the very small space which the width of the far end of the street fills in your eye; except the few which came straight at you, and they would appear to stand still ; and so do a few of the meteors. So they appear to radiate from that point of the meteoric ring, or the stars beyond it, toward which the earth is going just then, or whose longitude is 90° less than the sun's. The height of the meteoric shower will not occur until we have come in sight of that point in the sky toward which the earth is moving, *i.e.*, until those stars have risen; which must be near midnight, just as the moon rises at midnight when she is 90° behind the sun. It begins about 15 minutes sooner in the south hemisphere than here, because that part of the earth touches the inclined plane of the meteoric ring first.

All the phenomena of the November 'showers' appeared to suit Mr. Newton's hypothesis of a thin ring of meteors, with a longish lump in it, all revolving the opposite way to the earth in 354·62 days in a nearly circular orbit rather smaller than ours, inclined 17° to it and crossing it where the earth is on November 14 now;—34 such periods = 33 years; and so we fall in with the great shower every 33 years, and sometimes the next year also on account of the length of the lump, but in other years we cross the thin part only. The nodes recede with reference to the direction of the meteors, or advance in longitude, 52″·4 a year: which, with the precession of our equinoxes 51″·1, makes the meteors a day later nearly every 34 years (only an accidental concurrence with the other 34).

But Mr. Adams has calculated that the nodes of such an orbit would not recede 52″ a year under the disturbances of the planets, but only 21″; and that the orbit which does satisfy all the conditions makes the period of each meteor those same 33·25 years which have been the average period of the great shower ever since 13 October, 902 (= 18 Oct. N. S.), with an eccentricity 54 times greater than the earth's orbit, and a major axis 10·34 times greater, reaching therefore beyond Uranus and crossing our orbit near the meteoric perihelion. Consequently the thickest part of the ring crosses our orbit once in the 33·25 years instead of 34¼ times (though we meet it only once) according to the other theory. As the great shower lasts two hours and the ring is 17° inclined to the ecliptic, it must be about 35,000 miles thick. And from the necessary velocity in such an orbit that thick part of the ring must be above 800 million miles long near perihelion (p. 261), as we sometimes cross it two years running; and the whole orbit is 4420 million miles long. This is very like some of the comets' orbits ; and the August meteors are now thought to have a much longer orbit even than this. Astronomers have already identified the orbits of two known comets with these two meteoric rings.*

The meteors of all the meteoric days have much the same character. They are mostly very small, Mr. Herschel says few above a pound in weight, and many only a few grains. Their average height above the earth when we first see them appears to be about 75 miles, and they generally disappear at 50 miles, in consequence of being burnt up by the heat they generate by friction in passing through the air with a velocity of 2500 miles a minute. That is at once the cause of their brilliancy and the reason why so few of them are found; and those that are found bear marks of intense heat and of their outsides having been melted.

---

* See R. A. S. Monthly Notices of April, 1867.

Some of them may pass by the earth altogether and continue their orbit round the sun.

The number of meteors in one of the really great showers is enormous: the one of November, 1866, had only about 8000 at Greenwich, which is quite scanty compared with some that are recorded, in which 300,000 were said to be reckoned at the same place; and they are so near the earth that observers at places a good way apart must see a different set of meteors.

There is a different class of much larger stones called aërolites and fireballs, which come singly and at no fixed times, and generally in the afternoon, it is said, while the meteor showers are chiefly after midnight. Accounts of the doings of such stones in various ages of the world have been published lately. Not a few people have been killed by them, and buildings set on fire. Their roaring noise, like their heat, is only caused by their rushing through the air. They generally contain a great deal of iron and bismuth, and none of them any substance unknown upon the earth.

CHAPTER V.

WE will now return to the planets and their satellites, to consider the laws by which they all move, and may move for ever, by virtue of certain consequences of their laws of motion, and the small eccentricity of their orbits, which enables them to recover from every disturbance, and makes all the inequalities only periodical, and the equilibrium of the system stable.

Kepler's laws.—I have already mentioned one of those three laws which Kepler found by observation that all the planets follow, and which Newton afterward proved by mathematics that they must follow, under the universal law of gravitation. The one I mentioned was, that the planets move in ellipses round the sun, in one focus of the ellipse.

The second is that the radius vector from the sun to any planet sweeps over equal areas in equal times in the same orbit, though not in different orbits ; therefore going faster when the planet is near the sun than when it is further off, as I said about the moon and Mercury at pp. 115, 165. This is called the law of *conservation of areas ;* and it is only the same thing in other words as saying that the angular velocity in any given orbit varies inversely as the square of the distance.

For the area of any narrow sector of the orbit, be-

tween two radii so near together that we may consider
the bit of orbit a straight line, is half the numerical value
of the small angle between them × the product of the
two radii—or × the square of either of them, as they
do not sensibly differ when close together ; and as that
small angle described in an unit of time is the angular
velocity, and varies (as may be proved by mathematics)
inversely as the square of the radius vector, it follows
that the area described in an unit of time, and therefore in any time, is invariable in the same orbit ; but
not in different orbits, under different original forces of
projection.   In fact the central force disappears in the
calculation of the area, which is left depending only on
the original distance and velocity across the radius
vector.   Therefore the radial disturbing force (p. 149)
does not affect the areal velocity of the moon, though
it increases her distance and diminishes her angular
velocity.   But the tangential force, so far as it is not
balanced on opposite sides of the orbit, does affect the
areal velocity ; and its gradual decrease, with the increase in the sun's average distance, retards the moon
in the way discovered by Mr. Adams (p. 151).

Although this law of angular velocity looks so like
the law of gravitation in another form, it has nothing
at all to do with it, and would be equally true if the sun
attracted at the tenth power of his distance, or in any
other way.   Therefore finding that the planets follow
that law proves nothing more than that the sun is their
centre of attraction according to some law or other.
Kepler's two other laws do depend on the attraction
varying inversely as the square of the distance.

The third of them is, that if you take the distances of any two planets from the sun, and cube them, and square the number of days or hours in their periodic times, the two cubes will bear the same proportion to each other as the two squares. Thus, omitting fractions, the cube of Mercury's distance in millions of miles is $35^3 = 42,875$; and of Venus's $66^3 = 287,496$, which is about $6\frac{3}{4}$ times 42,875. Then the square of Mercury's period in days is $88^2 = 7744$, and of Venus's $225^2 = 50,625$, which again is $6\frac{3}{4}$ times 7744. And so you would find it with any of the other planets, or with Jupiter's or Saturn's moons.

Gravitation and Inertia.—We do not know at all what the force called gravitation is, by which all bodies attract each other, any more than we know what magnetism or electricity is, by which pieces of iron in a certain state attract each other still more visibly and strongly at short distances. What we do know is that that force is universal, and acts even on invisible gases, such as common air or steam, just as it does on either the planet or the metal Mercury, or makes an apple fall to the ground because its stalk has got too weak to resist the earth's attraction. If you squeeze more air into a bottle by an air pump it weighs heavier in a balance; and weight (of things on the earth) is their mass multiplied by the attraction of the earth. One thing is heavier than another because it contains more of what is called 'matter,' which only means the stuff that gravitation acts upon, and which by its inertia resists disturbance (see p. 28). For it is quite an independent law of nature, that the inertia of bodies, or

10*

their power to resist attraction, is exactly proportionate to their quantity of attraction; or that their mass measured by either test is the same. For anything we know, attraction *might* have been in proportion to the surface, or something else which does not vary as the inertia, as electrical attraction perhaps is: that certainly does not vary as the mass.

The air on the top of a mountain is notoriously thinner and lighter, or there is less of it in a cubic foot, or in your lungs at every breath, than at the bottom. Why? Because the air near the earth is squeezed closer by the weight of all the air above it, just as a great heap of feathers is denser at the bottom than at the top, and also by the greater attraction nearer to the centre of the earth. I have already said (p. 67) that half of all the mass or weight of the air is contained within the height of some of the highest mountains, and that it decreases in density as it gets higher, until it becomes so thin that its force of expansion as a gas is balanced by the attraction of the earth. For the force which resists gravity in the air, and prevents it from being all squeezed down flat upon the earth, is the strong disposition of all airs or gases to expand and keep themselves loose. It is that which makes gunpowder explode, when the solid powder is suddenly changed by heat into a gas which wants to fill 4000 times the space, and has strength enough to burst open anything that resists it. Water turned by heat into steam acts in just the same way, only it is done more gradually.

There can be no doubt that the theory of gravity

varying as the inverse square of the distance was suggested to Newton, and indeed to Kepler who guessed at it before him, by what takes place with light and heat, and generally with all emanations in rays or straight lines. For the rays spreading out in all directions will evidently cover a space twice as wide and twice as high, when they have got to twice the distance from the centre from which they radiate. And as the space covered by the rays is four times as large, it must be covered only one-fourth as thickly as an equal space at half the distance.

But though this analogy was enough to suggest the law, and make it worth trying with the prodigious labor which Newton bestowed upon it, to see if it would explain all the motions of the planets, the analogy is a long way from explaining the cause or action of the law itself. For you observe that this law of the effect of emanations varying inversely as the square of the distance depends upon the size and position of the body which is to receive them. If you turn a board obliquely to the fire it gets less heat, exactly as a circle is diminished into an ellipse in perspective; or if you double up the same board or a plate of metal into a narrower one, it will also get less heat in the same position. But gravitation cares nothing about shape or position: every bit of matter in the universe has *found out* every other bit of matter, and is now attracting it according to its distance. Of all this there is nothing more to say but that it is the universal law of nature. All matter was created under this general condition, just as the various kinds of matter, lead, iron, air, water, trees, corn, and all other things, whether simple or compound, were created with special conditions or laws of nature of their own, which they always follow without hesitation or variety.

But though nobody knows why all bodies attract each other, or what gravitation is, we know very well the laws by which it acts, viz., that the attraction of one body on another varies as the mass of the body which we treat for the time as the attracting one, and inversely as the square of the distance of their centres of gravity if they are far apart, whether they are spheres or not. The attraction which we deal with in astronomy is measured by the velocity which the attracting body produces in a second of time in the attracted one. But the proportion in which each *moves* the other absolutely, or with reference to their common centre of gravity, if they are both free to move, is the inverse proportion of their masses : that is, the earth moves the moon eighty times as much as the moon moves the earth ; and the common point between them, round which they both really turn, is their centre of gravity, which is eighty times nearer the earth's centre than the moon's, as I explained at page 86.

So all the planets really move the sun, but in an insensible degree, as the weight or mass of them all is only about a 700th of the sun, and omitting Jupiter, the sun is 2500 times as heavy as all the rest. If they all pulled together in the same line, the c. g. of the whole solar system would still fall within the sun, as that of the earth and moon is within the earth. And Kepler's law of times and distances, to be stated quite correctly, say for Mercury and Venus, requires a double rule of three sum, such as this :—The square of Venus's period is to the square of Mercury's as the cube of Venus's distance multiplied by mass of Sun and Mercury is to the cube of Mercury's distance multiplied by mass of Sun and Venus. We might omit the multiplication if Mercury and Venus were equal, which they are *not;* or if they were both insignificant compared with the sun, which they *are;* or even if their *difference* was insignificant compared with the sun, which it is still more. So Jupiter's moons are too small to affect their own periods sensibly, but our moon is not ; and so we have to ' reduce the earth to rest ' in order to calculate the moon's motions, as described at p. 98.

But it may occur to you to ask, why does not a thing which weighs a pound at a yard from the earth weigh only a quarter of a pound at two yards, if attraction varies as the inverse square of the distance ? Because, as I told you at page 29, the attraction of the earth on anything outside it does not depend on the thing's distance from the surface, but from the centre, where the whole mass of the earth may be supposed to be condensed for all purposes of attraction, except where its spheroidal shape has to be taken into account. And no distance that we can move to makes any sensible difference in the distance from the earth's centre, except by such delicate tests as the time of clock pendulums, as in the experiments for weighing the earth.

I will now try to explain, as far as it can be done without mathematics, how gravity keeps the planets in their orbits, instead of either letting them run away beyond the power of the sun to pull them back again, or gradually pulling them closer and closer, till they fall into the sun.

Besides the creation of the sun and planets and their moons, and of this law of gravity, one thing more must have been done for them, since they certainly could not do it of themselves. They must have been projected or set going, both turning on their axes with the velocity they do turn with, and moving forward with the velocity which enables them to resist the force which tries to drag them into the sun. I say moving *forward*, because the only motion which they needed to start with was projection forwards in a straight line *across* the direction in which the sun attracts them. The laws of motion could do all the rest. For that was enough to start the contest between the two forces which are always contending against each other—the force of original projection, under which they are always trying to run off in a straight line, and the force of attraction which is always drawing them away from that line, and bending their course toward the sun.

We talk of planets being 'projected,' 'set spinning,' and so forth, not because any astronomer believes they were suddenly started by the Creator in that way, or that their motions are not the result of some very simple original disposition of the matter of the universe and the laws ordained for it; but merely because the present condition of things is what it would have been if the planets had been so projected and set spinning under the known laws of nature.

THE LAWS OF MOTION, AND ELLIPTIC ORBITS.

The first of Newton's three fundamental laws of motion, on which the science of *dynamics* is founded (which word means the effect of force in producing motion), is that a body once in motion will go on for ever in the same straight line and with the same velocity, till some new force turns it aside. So the force of original projection keeps a planet always moving on across the line between it and the sun, and produces what is known as the centrifugal force; and attraction is the second force, which is always drawing it aside from a straight line into a curve. You see the same thing in a small way in every stone you throw : it wants to go forward in a straight line, by the first law of motion, but is constantly drawn out of it by the attraction of the earth ; and so it moves in a curve, and at last falls to the earth because the attraction is too strong for any velocity we can give it. If you throw a ball across the wind, the wind also bends its course into another curve sideways ; so there you have three forces acting, and producing a motion compounded of them all, in a ' curve of double curvature.'

The second law of motion explains the action of these double forces, or of any number of forces. That law is, that a second force draws a body aside from the straight line which it is taking under the action of the first, just as much in any given time as if it had not been moving at all, or had been moving with any other velocity. The consequence is, that at the end of the time it is found just where it would have been if the

first force had first sent it to its place in the first
straight line, and then the second had suddenly carried
it off from there in another straight line ; or conversely,
as if the first force had sent it forward in no time at
all, and the second had then moved it in its own time.
And that again comes to the same thing as if the body
moved along the diagonal of a parallelogram (or a four-
sided figure with its opposite sides parallel) of which
one side is the course it would have taken in the given
time under the first force alone, and either of the two
cross sides represents the course it would have taken
under the second force alone.   This is called the *composition* or the parallelogram of forces.   The effect of
any third force is added in the same way, taking the
diagonal which represents the *resultant* of two of the
three forces, to form one side of another parallelogram,
and the third force as the cross side; and so on.   But
here we have no occasion to consider more than two.

The truth of this second law of motion may be
shown by a machine which is so contrived as to shoot
one ball forward by a spring at the same time that it
drops another; and though one ball of course travels
much further than the other, you may hear them both
fall on the floor together.   So if you drop anything in a
railway carriage going forty miles an hour, it falls as fast
as it would if the carriage were standing still, though
it has really moved in a curve combining the forward
motion of the carriage with the downward motion under gravity.

Newton's third law is, 'Action and reaction are
equal and opposite ;' or the planets attract the sun as

much as he attracts them, though they do not move him as much. It has been proposed to substitute for this the following: 'The velocity generated in a given mass by any force in an unit of time is proportional to the force.' But that is really part of Newton's second law and has no relation to the third.

What are called '*moving forces*' are measured by the product of the mass moved × the velocity generated in the unit of time: which is a second. That product is also called *momentum*. Velocity means the rate at which the body would go on moving *uniformly* if no more force were applied to it, and is always measured in feet in calculations of forces.

This measure of moving force, or the force a body moves with (as if we were calculating the force with which the earth would strike the sun) differs from the measure of gravity or attraction at page 28, which disregards the mass of the body moved. That is the measure used in astronomy, and in most problems of motion; for we generally want to find the motion of a body in a given time, and not its momentum. Attraction *between* two bodies, measured either as a moving force, or as the pressure they would produce on a spring between them, is the product of the two masses, divided by the square of their distance; while the *accelerating force* of attraction is the mass of the body whose motion we are not considering, divided by the square of their distance. If we want to treat one of them as fixed, instead of the centre of gravity of the two, we must reckon the accelerating force on the other as the *sum* of the two masses divided by the square of their distance, as I explained at p. 98 about the earth and moon, and in all the calculations of periods.

The momentum of a moving mass is also called its 'quantity of motion.' But the *square* of the velocity × the mass, or more conveniently, × half the mass, is called its 'force of motion' or *vis viva ;* which is the measure of the work done or capable of being done by such a weight moving with such a velocity, before it is all used up. For that vis viva of a body falling may be proved to = its weight (or mass × the force of gravity) × the height it has fallen ; which would of course raise another equal weight through the same height, as a measure of the work done. Again the vis viva of rotation, as of a fly-wheel or of the earth in raising the tides (p.144) is half the mass of each particle × its own velocity^2, or × its distance^2 from the axis × the velocity^2 of rotation : which therefore depends on the shape and density as well as the weight ; being evidently greater the more the body spreads out from the axis of rotation.

And as 'forces of translation,' which merely drive a body along, are resisted by the inertia of its mass, so rotatory forces are resisted by a rotatory inertia called *moment of inertia,* which is the sum of the masses of each particle × its distance^2 from the axis. So that in each case the vis viva or force of motion is the proper inertia × velocity^2 of translation or rotation. That is why a large thin grindstone is both harder to drive and harder to stop than a small thick one of the same weight. The moment of inertia of a sphere or spheroid is $\frac{2}{5}$ of what it would be if its mass were all condensed into a thin band round its equator; and therefore varies as the square of the equatorial radius. The time of rotation of a shrinking globe also decreases as the square of that radius; because each particle obeys the law of conservation of areas (p.223). Therefore the day would shorten an hour if the earth contracted a 48th. But the day has certainly not shortened since the earliest recorded eclipses; and therefore the earth has not shrunk in that time at any rate.

If a stone could be thrown fast enough, the attraction of the earth would not be able to bring it *to* the earth, but only *toward* it, pulling it aside from the straight line in which it would go through space if there were no attraction ; and so it might go round the earth for ever, like the moon.

You know that centrifugal force is that which stretches the string by which anything is swung round. But the word centrifugal, or centre-avoiding, gives some people an incorrect idea of what that force is. It is not really a force which makes things fly away from any centre, but makes them go straight forward whenever they are set at liberty. If you let the stone go out of a sling when it is at the highest or the lowest part of its swing, it does not go straight up or down, but forward, in the direction of a tangent to the circle in which it was swinging. So what is called centrifugal force is really due to nothing but the desire the body has to move on in a straight line with the velocity which it has at each moment, according to the first law of motion: we shall calculate its amount presently. The force which draws a planet away from the straight

line in which it desires to fly off, brings in the second
law of motion, and is called the *centripetal*, or centre-
approaching force; which of course is the attraction
of the sun.

Therefore if a planet is ' projected,' or at any time
finds itself moving, exactly at right angles to its radius
vector, as a circle is everywhere at right angles to its
radius, and with a velocity producing a centrifugal force
which exactly balances the sun's attraction at that
distance, it will go on moving in that circle until some
other force comes to disturb it. If that had been so,
and if there were no disturbances instead of many, we
need say no more; and then we might never have
known what the law of gravitation is; for planets could
revolve in circles undisturbed under any conceivable
law of force. But their state of equilibrium might be
such that the smallest disturbance would either send
them into the sun, or else away from him for ever. We
know that none of the planets do move in circles, and
that their velocity and direction are never both together
such as a circular orbit would require; and therefore
we must consider what sort of orbit they will describe in
those circumstances.

Let us take the most simple case of a planet found
moving at any distance from the sun which does not
vary for a short time, but with a velocity too great for
a circular orbit of that size. Then the distance must
inevitably begin to increase very soon, just as a weight
swung round by an elastic string will stretch it and fly
further out if you increase the velocity. But will it
merely describe a larger circle, as the weight does?

No : because the force of the sun diminishes as it gets further off, but the elastic force of the string increases ; and if the velocity were too great for the attraction in a small circle, it will be still more too great for the weaker attraction in a larger circle. The next question is, will it go further and further off as the attraction gets weaker, and so describe a spiral widening out into infinity? That question cannot be answered without mathematics, beyond merely saying that it will not describe such a spiral under the existing law of gravitation, though it would under another not very different. We can however show in a general way how the planets escape running off into spirals, and describe ellipses instead, and how they manage to get away again from the sun after approaching nearer and nearer to him with his attraction increasing. We cannot get beyond this without mathematics ; but you will not be surprised at that when you know that Newton had to invent a new method of calculation before he could prove that the orbits must be elliptical.

Elliptic Orbits.—I have several times spoken of a force being 'resolved' into some direction oblique to itself. As two forces can be compounded into one, which is called their *resultant* (p. 232), so one force can be divided or resolved into two in any directions we please ; except that there can be no resolved part of any force in a direction at right angles to its action. That is the reason why we need not consider the resolution of forces in circular motion ; for the motion in a circle is always at right angles to the radius, or to the direction of the central force. But in an ellipse it is

not so, except at the two apses. If you look at this
figure of an ellipse, you see the motion of the planet
P in either direction,
toward T or *t* (both in
the tangent to the el-
lipse), is not perpen-
dicular to PS, the line
to the sun. So that
if it is going in the
direction PT, part of
the sun's force goes to
increase the velocity in
that direction, while
the other part is pulling it out of that direction. The
sun's force then may be resolved into two forces, one
in the tangential direction PT, and the other in the
direction TS perpendicular to PT; and these two forces
are proportionate to those two lines. On the other
hand, if the planet is going away from the sun toward
t, part of the sun's force resolved in the same way goes
to diminish the planet's velocity in the direction P*t*,
and the other part toward drawing it aside out of the
direction P*t*.

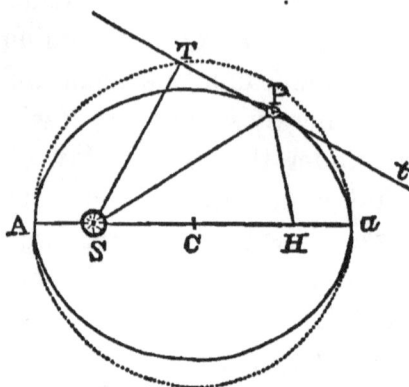

Then while it approaches the sun S, or while the
radius vector SP diminishes, the force represented by
PT continually augments the velocity, till at last at
some point A the velocity becomes so great that the
centrifugal force first balances and then beats the centripetal; and so that point becomes an apse, or a place
where the radius vector changes from decreasing to increasing, or *vice versâ*. So we may say that the sun

enables the planets to run past him at perihelion by
having made them run faster toward perihelion—only
not *straight* toward him. The converse of this takes
place when the planet is approaching aphelion, in a
direction like P*a* (but on the lower side of the figure,
or reversed). Then the sun has been gradually reducing the velocity by that part of his force which is resolved into the direction PT, now acting against the
increase of the radius vector. At last the velocity is
reduced so much that attraction prevails over centrifugal force, and the radius vector can increase no longer,
but begins to diminish again at *a* the further apse, and
then the planet begins to approach the sun again, and
so the orbit is completed. But why the apses recur at
the same places, under a force varying exactly as the
inverse square of the distance, and why the orbit is an
ellipse and not some irregular or egg-shaped oval, cannot possibly be explained without mathematics; and
even they will only prove that it is so.

I showed you in the moon's disturbances (p. 152),
that if the central force is less at aphelion than it would
be according to the square of the distance, that is if it
varies in a rather higher ratio, the apses do not recur
at the same places, but advance, and the orbit is a kind
of spiral returning into itself, that is, a revolving ellipse.
I now add that if the central force varied in as high a
ratio as the inverse cube of the distance (as disturbing
forces do), the centrifugal force and the force of attraction could never balance each other again if either of
them once got the best of it; and so the orbit would
become a complete spiral never returning into itself

again, and the planets would either run at last into the sun, or else further and further off into infinity, according to which force had accidentally preponderated for a moment.

The only other law of force of which we have any experience, and that only as an indirect result of the universal law of gravitation in some special cases, is when the force varies directly as the distance, as in the case of a pendulum bob which is attracted to the centre of its vibration by a force very nearly in proportion to the distance from it, whether the pendulum is vibrating in one plane, as in a clock, or revolving as a conical pendulum; and in a mass of stars, as we shall see afterward. It is singular that in that case also an ellipse is described, though the law of force is so extremely different; only the centre of force is then in the centre instead of the focus of the ellipse. The reason why the apses of the pendulum ellipse revolve, as stated at p.101, is that the force does not increase quite as fast as the distance; and so both apses are in the condition of the moon's apogee at p. 152.

Centrifugal Force.—We have so much to do with centrifugal force in these matters that you may as well see how it is measured. It is the force which resists deflection from a straight line into a circle, if the body moves in one, or into that circle which agrees for a moment with whatever curve the body moves in, and which is called the ' circle of curvature ' there; and it is measured by *twice* the amount of such deflection in the unit of time, which is a second: that being the velocity *per* second which such a force would produce *in* a second, according to calculations which you must take for granted. Now the deflection of any small arc from the tangent at the end of it = the square of its length divided by twice the radius of the circle, as I said of the roundness of the earth at p. 13. But that small arc represents the velocity of the body in its orbit. Therefore centrifugal force = the square of the linear velocity divided by the radius of curvature.

Again as linear velocity = radius × angular velocity, when the radius is not changing, centrifugal force also = the radius vector × square of angular velocity, in a circle, or near the apses of an ellipse. Centrifugal force is not a force at all in the proper sense of being a cause of motion, for it is only a consequence of motion ; but it is a very convenient measure of the force which balances it, and which truly is one, whether it is the tension of a string or the attraction of the sun ; for in either case it pulls the body out of the straight line in which it would move otherwise.

But you may ask, how is angular velocity of degrees and minutes to be reduced to common numbers? Simply by the fact that an arc of 180° is 3·1416 times the length of the radius ; and therefore the angle of 180° = 3·1416, which is usually denoted by the Greek letter π for shortness. Consequently 4° nearly = ·07, and 57° 17′ 45″ = 1. Thus the earth's angular velocity of rotation is expressed by 360° × 1·0027 (p. 58) or 6·3005, divided by 86,400, the seconds in a solar day. And the centrifugal force at the equator is the square of this angular velocity × the earth's radius of 3963 × 5280 feet. You will find this = ·11127 ; which is a

10

289th of the force of gravity at the equator, which would be 32·2 if it were not diminished by the centrifugal force to 32·09 (p. 246).

When the radius vector changes, linear velocity no longer bears such a simple relation to angular; but this rule holds always: whatever curve a body moves in, the linear velocity varies inversely as a perpendicular drawn from the sun, or the place which the radius vector is measured from, to a tangent of the curve at the point where the body is, or to the direction of its motion for the moment, in which it would go on if the attraction were suddenly withdrawn. The tangent TP*t* of an ellipse (p. 238) is easily drawn from the fact that it makes equal angles with the two focal distances SP, HP; and ST, the perpendicular to it, always cuts it in the circle which contains the ellipse. No planetary orbit is anything like so elliptical as I have made the figure for distinctness, and therefore ST, the perpendicular on the tangent, is really much nearer to SP, the radius vector. The linear velocity then varies inversely as ST, but the angular inversely as SP^2. And that is true whichever way the planet is going, whether toward T or *t*. At the apses the perpendicular ST coincides with the radius vector, and the former rule holds.

A planet then goes slower the further it is from the sun, but not so much slower in linear as in angular velocity. What I said before about the increase of velocity from aphelion to perihelion, and the decrease from perihelion to aphelion, referred to linear velocity in miles; for it is actual velocity or motion by which all forces are measured, and which the laws of motion govern. Therefore taking our example of Mercury again, its linear velocity at perihelion is greater than at aphelion in the proportion of 3 to 2, while the angular velocity is as 9 to 4 (p. 224). And the linear velocity is always made up of that due to the original impulse (which never wears out), increased and diminished from time to time by the resolved or tangential part of the attraction of the sun.

WEIGHING OF THE SUN AND MOON.

When I said that the sun is 316,560 times as heavy as the earth (p. 63), I promised to prove it before we finished the book. It can be done independently of the moon, and in another way through the moon (which is the usual one), by common arithmetic and nothing harder than a few sums in fractions; supposing you understand them, and remember that the same quantity may be divided out of a numerator and dedominator, or out of the numerator of one fraction and multiplied into the denominator of another which is equal to it, and that two equal fractions or sums with the mark = between them are called an *equation*, and that both sides of an equation may be squared, or have their square root taken, or be multiplied or divided or increased or diminished, by the same quantity, without affecting their equality.

We are now in a condition, first to weigh the sun against the earth; secondly, to prove the necessary truth of Kepler's law of the square of the periodic time varying as the cube of the distance; and thirdly, to carry it further, and show the relation of the time of any planet by itself to its own mean distance from the sun, which is a very different thing from comparing the times of two planets with each other. Kepler's law would have had no meaning at all if there had been no other planets than the earth, and would have been worth very little if there had been only two or three : indeed we have seen that Bode's law of distances holds (pretty nearly) from Venus up to Uranus, but fails both for Mercury and Neptune. But Newton's law of time and distance is as good for one planet as a hundred ; being a necessary consequence of the law of gravitation, that the attraction of a globe (or anything else if very distant) varies as the mass divided by the square of the distance from the centre.

As we must begin somewhere, with some proposition already proved by mathematics, let us start with this very simple one, which will also simplify our calculations considerably. The time of performing an elliptical orbit round the sun in the focus is independent of the eccentricity ; and therefore it is the same as if the eccentricity were nothing at all, or the ellipse a circle with the sun in the centre, and that circle the one which contains the ellipse. Therefore also in calculating the whole period of revolution we have no need to trouble ourselves about the variations of velocity in different parts of the orbit, and we may (for this purpose) treat the orbit of the earth or any other planet as not almost, but absolutely circular. .

As we are going to weigh the sun in earths we must remember that we know nothing of the earth's mass except through its power of attraction, as in the experiments for weighing it. And we must see how that is measured before we can go further. Like all other forces, it is measured by the velocity it generates in the agreed unit of time, a second : and a foot is the agreed unit of length, and therefore of velocity. By a machine called Attwood's (of which you may see a picture in the English Cyclopedia) an accelerating weight can be made to act for a second on one of two others, which are balanced on a string over a light pulley with as little friction as possible. After the accelerating weight is stopped or thrown off, the others are found to move on with a momentum equivalent to that of the accelerating weight moving with a uniform velocity of about 32 feet *per* second. The momentum of the accelerating weight shows that it would have fallen only 16 feet *in* the first second, but would have gone 64 feet in the second, 16 × 9 in the third, and so on, if it had been left falling without impediment. But independently of that, which involves mathematics, here you have the experimental fact that gravity generates a velocity of 32 feet per second ; which is therefore the figure by which it is always represented, subject to small corrections.

A more accurate way of finding it, free from the impediments of friction, but involving some mathematics, is by measuring the length of a pendulum which vibrates in a second. The length is not measured to the centre of gravity, but to a point a little lower, called the *centre of oscillation*, which depends on its shape, but can be got very accurately by a well-known contrivance, described in the Rudimentary Treatise on Clocks (p. 45, ed. 4). I can only tell you here that

gravity = the length of the one second pendulum ×
π^2 (π being the symbol commonly used for 3·1416).
That length in England is 3·26 feet (for we must reckon
in feet), and therefore gravity = 9·86 × 3·26 = 32·14.
But that is rather too little on account of the centrifugal
force of rotation (p.242), and gravity may be taken at
32·22 feet on the surface of the earth as a non-rotating
sphere with a radius of probably 3958·6 miles (p.24).

Now let us see what proportion the sun's force must
bear to that, in order to balance the centrifugal force
of the earth's revolution (not rotation). Centrifugal
force in a circle = linear velocity^2 divided by radius
(p. 241). The radius of the earth's orbit is 91,404,000
× 5280 feet: its velocity is very nearly 18·2 miles (p.
36), and the velocity^2 is 331·2 × 5280^2 feet. Therefore

$$\frac{\text{cent. force}}{\text{gravity}} = \frac{331 \cdot 2 \times 5280^2}{32.23 \times 91,404,000 \times 5280} = \frac{1}{1684 \cdot 2}.$$

'Cent. force' may stand for either centrifugal or centripetal, for they must evidently balance each other.
Therefore gravity is 1684 times the sun's attraction
here; and if the earth weighed 1684 tons and was
held by a chain instead of the sun's attraction, the
chain would only have to bear a strain of one ton.

Now for the next step, we know that the central
force is the sun's mass divided by his distance^2, and
that gravity is the earth's mass divided by its mean
radius^2, and that the latter is 1684·2 times the former.

$$\text{Or} \quad \frac{\text{earth}}{3958 \cdot 6^2} = \frac{1684 \cdot 2 \text{ sun}}{91,404,000^2}; \text{ therefore}$$

$$\frac{\text{sun}}{\text{earth}} = \frac{91,404,000^2}{3958 \cdot 4^2 \times 1684 \cdot 2} = 316,565;$$

which is very near indeed to the 316,560 which I gave as the result of other calculations. These two sums might have been done in one, but I wanted to show you the proportion between the sun's attraction here and gravity.

If you have a fancy to see the weight of the sun in tons, you must turn the earth into tons from its bulk in cubic miles (p. 24), and its density $= 5\frac{1}{2}$ times that of water, of which 36 cubic feet weigh a ton. Then multiply by 316,560, and you will find the tons in the sun to be about 1850 with 24 cyphers after it, or 1850 septillions, as the earth is 5842 with 18 cyphers after it, or 5842 quintillions of tons.

The following is the usual way of weighing the sun; but it requires the moon's weight to be known first to do it accurately. The mean radius of the earth's orbit is 382·76 times that of the moon round the earth, and the moon's mean angular velocity is 13·37 times the earth's, being inversely as their sidereal periods (p. 108). Therefore the moon's mean linear velocity round the earth reduced to rest (p. 98) is to the earth's round the sun, as 13·37 to 382·76. The sun's force on the earth is to the earth's force on the moon as their deflections from a straight line per second, which are as the squares of their velocities divided by the diameter of each orbit, as I explained for centrifugal force at p. 241; and those diameters are as 382·76 to 1. Therefore the deflections are as 382·76 to 13·37^2. But the sun must be 382·76^2 times heavier than if he were at the distance of the moon, to produce the effect he does here. Therefore

altogether $\dfrac{\text{sun}}{\text{earth} + \text{moon}} = \dfrac{382 \cdot 76^3}{13 \cdot 37^2} = 313{,}580,$

which makes the sun = 317,430 times the earth alone, adding one 81·5th for the weight of the moon.

But you see this result is wrong, exceeding the former one considerably, though it is near enough for a general explanation. Let us see if we cannot get a little nearer. I said at p. 148 that the sun's disturbance lengthens the moon's period as much as if the earth's mass were reduced a 358th. Therefore its real proportion to the sun is so much greater than it appears to be by calculation from the moon's actual period. And if you reduce 317,430 by a 358th, the result is 316,550; which is again very near the figure I have used throughout, deduced from the sun's mass as given in books of good authority for the old parallax (pp. 64, 86).

This same calculation evidently affords another mode of weighing the moon, if the sun has been first weighed independently; for the moon must be such a fraction of the earth's mass as will give the right result for the sun in the above calculation. In most popular astronomies the distinction between the earth and earth + moon is disregarded in weighing the sun; which makes a difference of nearly an 80th in his weight, besides misleading people as to the principle of measuring forces round a moveable centre, which I have several times mentioned. Moreover the following method of weighing the moon without the sun is founded on that very distinction.

E   G   M

It is not the earth's centre E, but G the centre of

gravity of earth and moon, that really describes what we call the earth's orbit. Therefore the earth is before its mean place at first half moon, and behind it at second half moon by the distance EG, which is now the thing we have to find, for we only knew it at p. 97 because we then assumed the moon's weight to be known. Now when Venus is at her nearest or 'inferior' conjunction, she is apparently displaced at half moons first one way and then the other, by a parallax or angle which = EG divided by the distance of Venus from us; and as that is known, and twice that angle is quite large enough to measure, being nearly 1′, therefore EG can be found. Then we know that the moon is to the earth as EG is to MG, and therefore the mass of the moon is found, as EM is known.

Mr. Airy said in 1856 that all the methods of weighing the moon make the earth 'about 80 times' her weight. If that were so then, this must make it about 83 now, since the distance of Venus from the earth is reduced a 24th, with all the planetary distances, and EG is increased in the same proportion while MG is not sensibly altered, being very much larger. You will see presently that another method gives the same result.

Law of Time and Distance.—We can easily prove also Kepler's third law independently, or rather that more complete form of it given by Newton, which determines the period of each planet from its own distance only. Centrifugal force = distance × angular velocity², still treating the orbit as circular (pp. 241, 244). Angular velocity per second is 360° or 2π or 6·2832 divided by the planet's period in seconds.

Therefore centrifugal force $= \dfrac{4\pi^2 \text{ distance}}{\text{period}^2}$. But this

must $=$ the central force or $\dfrac{\text{sun's mass}}{\text{distance}^2} : 4\pi^2 = 39\text{·}48.$

$$\text{Therefore period}^2 = \frac{39\text{·}48 \text{ distance}^3}{\text{sun's mass}}.$$

Strictly speaking, the denominator is the mass of the sun and planet together, for the reason given at p. 233. But as the sun is 1048 times heavier than the largest planet, we may practically disregard their weight, and consider the central force to be that of the sun's mass only.

But what is the meaning of the ' sun's mass ' in these fractions, and by what figures is it to be represented? If we say it is 316,560 times the earth's mass, we have only to put the same question about that. Generally we have to talk of masses only by comparison, and then we measure them by their three dimensions, of length, breadth, and height, and their density compared with water or the earth; or else we say that they contain as much matter as will weigh so many pounds or tons under the action of gravity here; or that the weight here $=$ the length × breadth × height × density × 32·2. But 'density' has no absolute meaning, only a relative one, and all these figures together do not tell us how much such a mass will move another in a second at the distance of a foot or a mile. We still want to know how to measure the attracting or 'accelerating' effect of any given mass; and we can only do that by referring it ultimately to the attractive force of the earth at the unit of distance from its centre, supposing it to be all condensed there; which we can measure easily.

For this absolute force of the earth (as it is sometimes called) is evidently $3958 \cdot 6^2 \times 5280^2 \times 32 \cdot 22$. Or the earth's mass at the unit of distance has 14,076 trillions of those things of which gravity is $32 \cdot 22$. Consequently the sun has 316,560 times as much, or has nearly 4456 quintillions of units of force at the unit of distance, which (remember) is a foot.

Weighing the Moon Independently.—We can now also weigh the moon without the aid of Venus, or the tides, or the still more delicate measurement of nutation, from nothing but her period and distance and the known measure of gravity. For the mass which determines her period is that of earth + moon: which must there-fore $= 39 \cdot 48 \times$ the cube of her mean distance, divided by the square of the number of seconds in her sidereal period of $655 \cdot 716$ hours, with a 717th taken off for the addition of a 716th by the sun's disturbance (p. 148). You will find that gives the absolute mass of

$$\text{earth} + \text{moon} = \frac{39 \cdot 48 \times 238,830^3 \times 5280^3}{2,357,300^2} =$$

14,246 trillions. But we saw just now that the absolute mass of the earth alone is 14,076 trillions, or 170 less than this earth + moon. The difference, or mass of the moon alone, is ·012 of the earth's mass; which comes very near to Mr. Adams's proportion of ·0123, and still nearer to the present result of the weighing of the moon by Venus at p. 248; for this makes the earth $82 \cdot 8$ times the moon; and as I said at p. 135, other authorities are inclined to make it rather more.

It seems odd that we should be less certain of the weight of the moon than of the sun and Jupiter, which are nearly 400 and 2000 times further off. But it is easier to find the mass of a heavy body from its attraction on light ones than of a light one from its disturbance of heavy ones.

You may think however that we have been only reasoning in a circle, because we have reduced the moon's period to meet its increase by the sun's disturbance, and yet that calculation of the disturbing forces at p. 147 assumed the moon's weight to be known. But there is nothing in that objection; for if you omit the moon's mass altogether in calculating the disturbances, putting 316,560 instead of 312,720 for the sun's mass, it makes a quite insignificant difference in this result. I should mention that I have not also reduced the moon's distance, though it is increased a little by the sun's disturbance (p. 148), because the increase of the period by a 716th is on the assumption of the distance not being increased; if that is taken into account, the period is increased (and must therefore now be diminished) rather more, and so the result would be the same, both here and at p. 248.*

The mass of Jupiter, or any planet with satellites, is found from the distance and period of any of them, by a calculation like that which I gave just now. For Jupiter's absolute mass, or force at the unit of distance, = 39·48 × the cube of any moon's distance (in feet) divided by the square of its period (in seconds), which you will find in the table at the end. And then di-

---

* See Archdeacon Pratt's Mechanical Philosophy, § 299.

viding the result by the 14076 trillions for the earth's mass, you have the proportion of Jupiter to earth.

Or the mass of Jupiter may be compared with the sun, through any of his moons, as we compared the sun with the earth through our moon ; except that Jupiter is too heavy to be neglected in comparison with the sun, and his moons may be neglected in comparison with him—just the converse of the earth. The result is,

$$\frac{\text{Sun}}{\text{Jupiter}} = \frac{\text{♃ \& ☉'s distance}^3 \times \text{☾'s period}^2}{\text{♃ \& ☾'s distance}^3 \times \text{♃'s period}^2} - 1 ;$$

the 1 meaning only that Jupiter's mass is not neglected. This ought to come 1248 or nearly so, and 302 for the proportion of Jupiter to the earth, as stated at p. 171. For Jupiter's mass has been also verified several times from his disturbances of some of the asteroids, and all the calculations give nearly that result.

The distances of Jupiter's moons are given in all the books only in terms of his equatorial radius, as in the third column of the table at the end. But those reduced to miles, by the diameter which is generally given, all produce a mass about a 17th too great, by either of the calculations ; therefore the distances are nearly a 50th too great : which is far too large an error to be accounted for by the mutual disturbances of the satellites or by Jupiter's oblateness (p. 197). I have therefore given such distances in miles as will agree with this ascertained mass, neglecting the disturbances (as they affect the distances very little) and leaving the proportionate distances as I find them in other books ; observing only that astronomers are not quite agreed

about the proportion of Jupiter's diameters, and therefore about the length of one of them. It is not worth
while to go through the same process for Saturn's
moons; and Uranus's are too uncertain yet to say
more about them than I have at p. 194.

All these satellites are too small compared with their
primaries to be weighed through their own periods by
the method given for our moon just now. The masses
have to be found from their mutual disturbances. And
so the planets without moons have to be weighed from
their disturbances of each other, and of comets. All
that belongs to very high mathematics; but I have now
shown you how the dimensions and weights of all the
solar system are measured ultimately by a foot rule and
the vibrations of a pendulum.

And you can now appreciate the delicacy of the
Cavendish experiment for weighing the earth (p. 30),
though it looks easy at first sight to try how much one
mass attracts another. If you divide the weight of the
earth, 5842 quintillion tons (p. 247), by its absolute
force, you will find that it takes a mass of 415,000 tons
to exert a force of 1 foot per second at the distance of a
foot from its centre. Such a globe made of platinum,
the heaviest thing there is, would be 110 feet in diameter; and therefore you could not put anything to be
attracted by it nearer to its centre than 56 feet; and
its force at that distance would be only the 3136th of a
foot per second. Or to illustrate it by what is called
statical force, as in the Schehallien experiment, suppose
such a globe to be brought within a foot of the bob of
a long pendulum (hung, not balanced as in Cavendish's

experiment), it would only draw the pendulum out of the vertical by the angle whose numerical value is $\dfrac{1}{3136 \times 32 \cdot 2}$; or such a pendulum would have to be 1·6 miles long for its bob to be moved an inch by the attraction of that mass of platinum put close to it. So that the force of attraction, which produces such vast effects throughout the universe, is almost immeasurably small, except upon a very great scale.

Laws of Stability.—I will finish this account of the laws of motion of the solar system by stating three remarkable results of that permanent stability or equilibrium of the system which I have several times mentioned. The first is, that if you multiply the mass of each planet (taking the earth, or any other, for the unit) by the square root of its mean distance from the sun, and by the square of the eccentricity of its orbit, the sum of all those products is invariable, however each eccentricity may vary. The second is, that if you multiply each mass by the square root of its mean distance, and by the square of the numerical value (p. 241) of the inclination* of its orbit to a certain fixed plane near the ecliptic, the sum of those products also is constant.

The third is the most striking of all. The sum of all the areas swept over by the radius vector of every planet, satellite, and comet in the whole system, in any given time, is invariable, though they all may be made to deviate a little from the law of the conservation of

---

* Strictly it is the trigonometrical *tangent* of the inclination, but the angles are too small for the distinction to be important.

areas (p. 223) by their mutual disturbances. The area so described by any planet in a second, being constant in the same orbit, is the area of the whole orbit divided by the number of seconds in its year; and that area is 3·1416 × the product of the two semi-axes, or × the square of the mean distance if the orbit may be treated as a circle. From this, and the law that the square of the period varies as the cube of the distance, you may easily deduce another, that the areas described in different orbits in any time round the same central mass are as the square roots of the distances. Therefore the *areal velocity* of distant planets is greater than of near ones, though both their angular and linear velocity is less.

THE FOUR CONIC SECTIONS.

There are two other curves beside circles and ellipses, in which the planets could, and some of the comets do move, under the law of gravity varying inversely as the square of the distance. They are the *parabola* and the *hyperbola*. A parabola is the curve described by a stone thrown into the air any way but straight up, or by water rushing out of a hole in the side of a pretty full cask, except that the resistance of the air spoils the accuracy of the curve. You may see a hyperbola in the shadow on the wall from a round shade over a lamp set near the wall. But the remarkable thing is that these four curves, the circle, the ellipse, the parabola, and the hyperbola, which are the only possible ones* for the heavenly bodies to describe under the existing laws of nature, are all produced by cutting straight

---

* Appendix, Note XXII.

through a cone in different directions. You may have heard the term *conic sections* without knowing what it means, or why whole books of mathematics are written about them. There is no part of geometry with so many curious and elegant problems in it; and the reason of its importance is that all the heavenly motions are performed in conic sections. I must therefore explain what they are, as far as I can without mathematics.

You probably know that a cone is a round pyramid, or a thing with a circle for its bottom and a point for its top, and straight sides. When the top is vertically over the centre of the circle it is called a *right cone*, and when it is not, an *oblique* one. Consequently the outline or any vertical section of a right cone is a triangle with two equal sides. If you cut a few cones out of some soft stuff such as a turnip, or get them turned in wood to be sawn through afterward, you will be able to get a better idea of the different conic sections than from any pictures, and therefore I shall not give any. First of all it requires no cutting to see that all sections parallel to the base of the cone, or at right angles to its axis, must be circles; and a circle may be considered an ellipse of no eccentricity, or with its two foci run together. Next, any other section of the cone which comes out at both sides is an ellipse, as you will see by cutting it through in that way. And I should tell you that the cone is considered to be extended to any length required for the elliptical section to come out again lower down on one side than the other. You would hardly believe without trying it, that the lower end of the section which comes out where the cone is broad is no broader than the higher end where it is narrow; and yet so it is.

An ellipse is also produced by any oblique section of a cylinder, as a circle is the section directly across a cylinder. But in fact a cylinder is only a cone of infinite length, so that its sides may be considered parallel, as when we look at the stars, which are some of them certainly much bigger than the sun, and yet look like points. This is the explanation of what I said at p. 40, that the oblique view of a circle is an ellipse, whether near or far off: if near, the lines of sight form a cone with the eye at the point; if very far off, they may be considered to form a thin cylinder.

A parabola is a very different looking thing, though it is produced by a very little deviation from the section of a cone which makes an ellipse. It cannot be cut out of a cylinder. It is like half—or rather, some indefinite part less than half, of an ellipse of infinite length, or one whose eccentricity $= 1$, or whose centre is so far from its foci that SC or HC (p. 39) may be considered $= $ AC. It has no minor axis, for its centre is at an infinite distance, that is, nowhere. It is made by cutting through a cone anywhere by a cut parallel to the opposite 'slant side.' Therefore there can be only one parabola, as there is only one circle, at any given distance from the top of the cone, which must now be considered of infinite extent downward. The two legs of a parabola are always getting more parallel to its axis, but never become quite so, however far they are extended.

A hyperbola is made by any section of the cone which is neither parallel to its slant side nor comes out of it at both sides. It does not differ much in appearance from a parabola, only spreading out wider. The slightest deviation of the parabolic cut toward the vertical turns the parabola into a hyperbola; but there may be any number of hyperbolas at any point in the cone. In some popular astronomies a vertical section only is said to be a hyperbola, but that is a mistake. Its legs continually approach two straight lines called *asymptotes*, which are in fact the outline of the cone itself, but never reach them, because they approach with a continually decreasing curvature: just as a series of weights of a pound, half a pound, a quarter, and so on, approaches 0, and their sum approaches two pounds, but the limit is never reached in either case.

A parabola and hyperbola can theoretically be described by strings from the focus or foci (for a complete hyperbola has two, being two hyperbolas set back to back), but not practically. The focus of all the four conic sections is the point where they are touched by a sphere which also touches the hollow cone all round, like a ball put into a wine glass: if you draw the outline of the cone and the axis of any conic section such a sphere is represented by a circle.

The only thing I have to say about oblique cones is to reconcile what is said about the shape of countries being correctly represented in a stereographic map (p. 20) with the fact that every perspective view of a circle is an ellipse; and one perspective view of every ellipse is a circle. The explanation is, that an oblique cone allows two sets of circular sections, leaning opposite ways with respect to the axis of the cone; and in the stereographic projections of a small circle of the earth upon the plane of the equator seen from one of the poles, the oblique cone from the eye to that circle cuts the equator or any parallel plane in a smaller circle turned the other way; and therefore the shapes of countries in such a map are truly represented, though some of them must be on a larger scale than others in any flat map of a round world.

COMETS.

These singular bodies in several respects lie midway between the planets and the stars and nebulæ not resolvable into stars. For they have orbits with the sun in their focus, like the planets, and yet some of them probably run off among the stars, though they must take millions of years to get there. But they are not solid like the planets, being thin enough to see through in their densest part, and therefore nebulous, or composed of some kind of gaseous matter whose expansive force overcomes its mutual attraction and prevents it from coalescing into a solid lump. Most of them describe ellipses far more eccentric than the planets; but some describe parabolas and even hyperbolas, and those will probably never* pay another visit to the solar system through all eternity, but will run off till they are caught by the attraction of some other sun among the stars. The great majority of comets move in the same

---

* Appendix, Note XXIII.

direction as all the planets, and in orbits not very oblique to the ecliptic. The periods seem to be divided
into two classes, one from 3 to 7 years and the other
mostly above 70 years.

They take their name, which means hairy, from their
tails, though many comets have no tails, and to those
which have it is only a temporary appendage shot out
in the direction from the sun, when they get within a
certain—or rather uncertain distance from perihelion,
for they vary in that respect also. The head of a
comet is a round, bright, transparent, nebulous kind
of mass, sometimes much larger than any of the planets, and even half as wide as the sun, but so thin that
small stars can be seen right through the nucleus or
bright spot in the middle of the head; and so light
that they pass among the satellites of Jupiter without
disturbing one of them, though such comets have been
themselves completely thrown out of their orbits by
Jupiter's attraction. Their less violent disturbances
can be calculated beforehand, and their motions predicted accordingly, like those of the planets and the
moon, when their regular orbit has been once determined, as it can be by as few as three observations, fixing their velocity as well as their places, or by only
two observations at perihelion, the sun being always in
the focus of the orbit. Consequently there is not the
smallest doubt that they perform their journey round
the sun in obedience to the universal law of gravitation, however little matter they may have for it to
act on.

The greatest or perihelion velocity, either of planets

or comets, follows a simple rule according to the nature of the orbit. In a parabola, velocity[2] at perihelion $= 39\cdot48 \times$ sun's absolute mass (p. 252) divided by perihelion distance and multiplied by 2. In an ellipse it is that same quantity multiplied by $1 +$ the eccentricity, which is necessarily less than the 2 of the parabola. In a hyperbola the rule is the same as in the ellipse, but the eccentricity must exceed 1 to make a hyperbola; which cannot be explained without mathematics. A circle is only an ellipse whose eccentricity $= 0$. Therefore the velocity in a parabolic orbit is $\sqrt{2}$ or $1\cdot41$ times greater than in an elliptic orbit of very small eccentricity, at the same perihelion distance.

Consequently the whole orbit of a comet can be laid down in the map of space from only two observations at perihelion, with their exact times. For they give its least distance and velocity, and that gives the eccentricity, and the least distance and eccentricity give the mean distance (see p. 39); and the plane of the orbit must go through the two observed places and the middle of the sun, and that gives its inclination to the ecliptic and the longitude of its nodes. But though the linear velocity is least in a circle for a given *least* distance, it is greatest for a given *mean* distance; for a circular orbit is performed in the same time as any ellipse which it contains (p. 244), and which of course is a shorter orbit.

About 36 comets have been recognized as returning; some in moderately short orbits, others in very long ones, corresponding to their periods, according to the same rule of time and distance which all bodies going in elliptic orbits round the sun must follow. The comet of 1843 went nearest to the sun of any that is known, within 80,000 miles of his surface, or within one-third of the moon's distance from the earth. It swung past him with a velocity of 366 miles in a second, or 20 times as fast as the earth moves, and 1000 times as fast as any cannon ball, and was exposed to a heat four times as great as the melting heat of agates and rock crystal, the most refractory of all things, and yet came away none the worse, as far as we could see, being already in a state of gas. And it was bright enough to be seen in full sunshine in America.

It is curious that the grandest comet ever known, which shot out a tail of 60 million miles in two days,* and twice as long afterward, appeared in 1680, six years before Newton published the Principia. That also went much nearer to the sun than the moon is to us, and travelled 20,000 miles a minute, and was exposed to a heat 25,600 times as great as the hottest sunshine in the tropics. The earth only escaped a collision with Biela's comet by a month in 1832; so it is pleasant to reflect on their tenuity. That same comet suddenly split into two, which travelled amicably together and reappeared so in 1852. Since that time they have not been seen. Halley's comet, which followed Newton's by two years, and has a period of nearly 76 years, subject to delays by Jupiter and Saturn which can be calculated, in 1835 swelled to at least 74 times its size in 17 days after perihelion, till it

---

* Appendix, Note XXIV.

got too thin for its light to be seen, and so vanished; but no doubt it gathered itself together again, and will reappear in due time.

Comets seem to have a habit of apparently shrinking near the sun; but Sir J. Herschel thinks that that is really swelling so much that their outer parts become dissipated, or made too thin to be visible, as the whole comet was in 1835. I have already spoken of the gradual acceleration of Encke's comet, which returns in $3\frac{1}{3}$ years, and the theory founded upon it, at p. 83. Mercury's mass was calculated from the disturbance which it caused in this comet. The comet with the finest tail of modern times, was that called Donati's, which most of us saw in the autumn of 1858, and which nobody will see again for 2100 years, as it has to go 80 times as far off as Neptune. Its tail was 30 million miles long, and 90,000 wide just behind the head, and much wider beyond. You may see a fuller account of the strange behavior and peculiarities of the most famous comets in Sir J. Herschel's Astronomy, and in his ' Familiar Lectures.'

The tails are generally like a long brush in shape, but some spread out short and wide, like a 'bat's wing' burner gas flame: the story of one with 6 tails in 1744 is perhaps fabulous. Some tails are short and thin. But very nearly all of them agree in being directed from the sun, and projected with such velocity that it is evident that they differ from all other matter in being violently repelled instead of attracted by the sun, though not by their own matter or that of the comet whose attraction holds them together, until they are first exploded by the sun's heat and then projected backward by his repulsive force. The first explosion of tail matter does not take place backward, but rather forward, or toward the sun; and the tail is a little open just behind the comet, as it would be if the matter was shot out forward and sideways, and at the same time blown away backward by a strong wind or force from the sun. The apparent wheeling of a comet's tail round it, so as always to keep pointing from the sun, is to be explained like the appearance of the smoke or steam from a locomotive chimney, which seems to be carried along with it of a nearly constant length, and yet is really continually dissipated and becoming invisible at the far end, and continually renewed by emission from the chimney. A comet is continually shedding the far end of its tail by dissipation, and growing it afresh by emission.

Again the bend in the tail away from the direction of motion is due to the farther parts of the tail-smoke being, as it were, left behind by the near ones and by the head. For in order to keep pace apparently with the comet, the smoke ought to have a greater velocity forward when it has reached a larger orbit, than it started with: which it has no means of acquiring. Probably the tail-smoke is lost in space, and never gathered in again, as there is reason to believe the mere expansion of the head is. The nucleus, or bright spot in the middle of the head, generally remains visible, though stars can be seen through it.

It appears to be proved by two distinct methods that comets' tails are not self-luminous, though the heads

12

are. *Polarization* of light cannot be explained farther here than by saying that light reflected at an angle of 33° from a non-metallic surface, such as glass with a black cloth behind it, is polarized, or made to vibrate (see p. 78) in one plane cnly, and will not reflect again from another similar mirror set directly across the first, but is all stopped as completely as if the second glass were itself a black cloth. The light of comets' tails is found to be polarized, and therefore is concluded to be reflected light.

Mr. Huggins and Professor Miller have gone a step farther by the spectrum analysis (p. 68) and have ascertained that the light reflected is not of the same kind as would be reflected from the nucleus, which *is* self-luminous by the same tests; and therefore the tail-light must be reflected from the sun. The spectrum of an inflamed gas is not a continuous band of colors, but consists of a few bright lines at intervals; and the nucleus of a comet, as of some of the fixed nebulæ, appears to be only such a gas. There have been however very few opportunities yet of examining comets since this great invention of the spectrum analysis; and it is possible they may not be all alike. The extraordinary brightness of the comet of 1843 while very near the sun looks like the sun having at least a share in the effect.*

THE NEBULAR HYPOTHESIS.

Astronomers have naturally speculated what was the rudest and simplest state of the matter of the universe from which the present condition of the solar

---

* Appendix, Note XXV.

system may have been evolved by the known laws of nature and one original impulse from the hand of God. No one can reflect on all the motions of the system (except of a few moons and the November meteors) taking place in one direction, both of revolution and rotation, and all very nearly in one plane, without feeling inclined to attribute them to some common cause. And when we see, besides, the multitude of complicated motions, such as those of the moon, which are all unquestionably due to one law of gravitation, the probability becomes still greater of some very simple origin of the present state of things.

Gravitation is quite incompetent by itself, either to set a single body in revolution or rotation, or to determine the directions in which they should move—except to run all into a heap together. For anything that gravitation could do to prevent it, the 8 large planets and the 100 little ones, to say nothing of all the moons and rings, might divide the whole 180° of possible inclinations equally or unequally among them, and any number of them might go in opposite directions to the others. The chances against their all coming within such narrow limits and in the same direction by what is called accidental coincidence have been calculated to be above a trillion to one. They must either have been all deliberately put down where they are, and projected with the velocities they have, or else their motions must be the result of some much simpler state of things; and if that is possible under the existing laws of nature, nobody can doubt that it is more probable than the other alternative.

Laplace and others have come to the conclusion that it is possible; and that if an enormous globe of nebulous matter were once set slowly whirling in the general direction, which we call from west to east, it would gradually contract by cooling, and revolve faster (p. 235) and then throw off rings, and then each ring might break and gather itself into a globe, which would partake of the original rotation besides revolving in the general direction; and each globe itself might afterward throw off smaller rings, which would either stay as such, like Saturn's, or run together into moons.

The distances of the planets therefore represent the force of expansion of the original nebula against the attraction of its parts. Indeed whether this nebular theory is true or not, the distance of everything from everything else represents all the force which has been expended from 'the beginning in separating' them, or the vis viva they would acquire in coming together by attraction. And its increase as they approach is no 'creation of force,' as has been imagined; nor its decrease as they recede, a 'destruction of force:' it only re-appears in one case and becomes latent in the other.

The explanation devised for the abnormal motions of the moons of Uranus and Neptune is that those planets had their axes turned out of the perpendicular to their orbits before they threw off the rings which became their moons. But I have seen no suggestion to explain how the earth's axis could be so disturbed after it had become a rotating globe (which resists such disturbance) and thrown off a moon; or how the meteors come to revolve in the opposite direction.

It must be added however that experiments can be made to exhibit the process, and to show that a mass of fluid, and therefore of nebulous matter, set rotating, will throw off rings when its velocity is increased enough to overcome its cohesion; and that if the ring breaks it will gather itself up into a globe by the attraction of its particles. Plateau's experiment is this: water is lightened with spirits of wine till it has such a specific gravity that oil will lie anywhere within it. Some oil is put in, and immediately becomes a globe by its own attraction of cohesion (p. 18). The whole inclosed in a glass box is set spinning, with an axis through the oil to make it spin too: the oil globe first spreads into an oblate spheroid: then as it is whirled faster it throws off a ring, which revolves round it; after a time the ring breaks and gathers itself up into a smaller globe, which rotates besides revolving round the large globe; and then another ring is thrown off as the velocity is again increased, and so on.

The experiment is defective in neither being able to represent the attraction nor the contraction of the globe, on which the whole theory depends, for the reasons given at p. 169. The ring is thrown off when the centrifugal force exceeds the equatorial attraction, which decreases as the oblateness increases (p. 37); and that depends also on the proportions of inside and outside density. English mathematicians seem to consider this theory too uncertain to be worth expounding yet; but French ones have calculated from the present sun's rotation, that when it was as wide as each planet's orbit, it would turn in nearly the present period of that planet: and the same of the planets and their moons.

And it is easy to see that if the solar mass did turn at that rate, there would be the same equilibrium as now between the centrifugal and attractive forces; for the attraction on any planet, and therefore on the sun's equator, if it swelled into a globe as wide as the planet's orbit, would be the same as it is now (p. 29); and we shall see at p. 275 that a nebulous globe of any uniform density can all rotate together. But if the density then increases inward, the ' moment of inertia ' decreases, and the outside will turn faster and fly farther out, increasing the oblateness and perhaps separating as a ring.

An impulse given to a planet a little beyond its centre would produce both the present motions of rotation and translation, as you may see by striking a floating ball in that way; but it is difficult to see where such an impulse was to come from, acting in the same direction on all the planets.

NEBULÆ.

Bright patches in the sky, like a shining mist of definite size, are called nebulæ, whether they are large or small, resolvable into stars by telescopes or not. The success of Lord Rosse's great telescope in resolving into stars some nebulæ which had resisted all others, led a good many people to jump to the conclusion that all nebulæ consists of stars, and are only not resolved because they are too remote for the power of any telescopes yet made ; just as a distant flight of birds, or a plague of locusts in the East, which looks like a black cloud, can be distinguished either by coming nearer or being looked at through a telescope which magnifies the spaces between them. But we shall see reasons presently for rejecting this conclusion, and accepting the other, that some nebulæ are mere masses of stars, and that others are as truly nebulous as comets or their tails.

The grandest of all the nebulæ, the Milky Way, as it was named by the Greeks and Romans, is that band of strongish light which stretches over the whole sky in clear nights without a moon. That is resolvable into stars ; and our sun with all his planets is only one of them. For astronomers have come to the conclusion that this band of light is nothing but a mass of stars laid thicker together than they are in the rest of space in the form of a great grindstone, split and opened out in on ⁚ direction into something live the shape of a Y. If we were placed in the middle of such a mass, it would evidently look brightest in the direction in which there are most stars ; as when you stand in a narrow plantation of fir trees, it looks darkest in the direction of its length, because the trees are dark while the stars are light. And as they are all so distant that they appear equidistant, they look like a bright band all round us, of the width of the ' grindstone,' and as far off as all the stars appear to us.

The next most remarkable nebulæ of the same kind, but a vast deal smaller, are the things called the Magellanic Clouds in the southern hemisphere, though they are hardly nebulæ exclusively. Sir. J. Herschel, who erected an observatory at the Cape of Good Hope in order to complete his father's survey of the northern hemisphere of stars by adding to it the southern, describes these ' clouds ' as consisting of visible stars, of nebulæ resolvable into stars, and of nebulæ irresolvable, though for convenience the whole of each mass may be called a nebula. The larger of the two is about 6° wide, or has 12 times the apparent diameter of the moon, or 144 times her apparent size. And from this is deduced a remarkable argument against the hypothesis of all nebulæ consisting of stars too far off to be distinguished. Unless the clouds are like pillars or very long ellipsoids turned endways to us, or if the depth is anything like the same as the width, the depth of the largest one corresponds to an apparent width of 6°, which means a width of about one-tenth of its distance from us; therefore the back parts are only so much farther off than the front. But it contains visible stars as well as nebulæ ; and a nebula of stars irresolvable by such telescopes as were used, must be, not one-tenth farther off than the visible stars, but many thousand times as far : therefore those nebulæ cannot consist of stars—if the visible stars are really in the nebulæ, and not in front of it, and therefore only optically within it.

Another argument for the existence of purely nebulous nebulæ is founded on the spiral shape of some of them, and on that shape coming out not less but more distinctly with increased telescopic power. Such a shape, like that of a bunch or mop of long thin feathers

or wisps whirled round, necessarily implies rotation of
a mass with some kind of continuity in it, almost as
certainly as if we saw it revolving, though it revolves
too slowly to have been yet observed; and their size is
probably enormous, and their density very small indeed; which, I will show you presently, implies a very
slow rotation. If the 'nebular hypothesis' is right,
these spiral nebulæ may be other solar systems in the
course of formation.*

Here again the great modern test of the spectrum
analysis comes in with a decisive answer to the question
whether all the nebulæ consist of stars, as some people
too hastily concluded when some new ones were resolved. Mr. Huggins and Professor Miller have found
about one-third of the 60 nebulæ they have examined,
to be merely gaseous like the heads of comets (p. 261);
and have even gone so far as to ascertain that nitrogen
gas is the principal element in the composition of some
of them. Other nebulæ give a continuous spectrum,
like the sun or a solid white hot body, and some have
an opaque nucleus within a gaseous envelope.

Some of them have other curious shapes implying a
definite constitution, such as the Crab and the Dumb-
bell nebulæ; and several have two heads like the divided comet of Biela and like double stars, each of the

---

* I take these arguments from the 'Plurality of Worlds,' a book well known
to have been written by the author of the Bridgewater Treatise on Astronomy,
and which (whether right or wrong) was very superior to the answers to it;
though of course the whole subject is one of speculation. It is now almost
certain that the time the earth has been occupied *by men* is a mere drop in the
ocean of its history as a planet; and the 'Plurality' only dealt with men as inhabitants of other worlds. It did not notice the great variations of heat in the
same planet described at p. 172.

12*

heads being manifestly a kind of focus of the nebulous matter which extends over them both. Some small round ones, of uniform density or brightness, are called *planetary* nebulæ, a very inappropriate name, it seems to me; for they do not wander, which 'planetary' means: they are certainly self-luminous, which planets are not: they are (as yet) unresolvable, and certainly not solid; and if they should turn out to be resolvable into 'star dust,' then they are still less like planets: their only resemblance is that they are round; for the smallness is only the effect of their distance. Though the largest of them has an apparent diameter not much greater than Neptune's, it has no visible parallax, which means (for stars and nebulæ) that the diameter of the earth's orbit is as nothing compared with its distance. If it had as much parallax as the nearest star, that apparent diameter implies a real one nearly twice as great as that of Neptune's orbit. Sir. J. Herschel says it must be far less brilliant than the sun, as one might expect of a nebula. A few nebulæ are annular, and one of those has been resolved into stars, as Saturn's ring is probably resolvable into a vast number of small satellites.

It is worth while to notice a remarkable result of the law of gravitation on the separate members of a large and uniform mass of either stars or nebulous matter. I said at p. 29 that the shell composed of the outer parts of a globe exerts no attraction one way more than another upon those within the shell. We proved at p. 250 that the square of the time of revolution varies as the cube of the distance divided by the mass of the attracting sphere. But if the sphere is uniformly dense or full of stars or nebulous matter, the mass itself varies as the cube of its radius, or distance of the outer particles from the centre. Consequently the time of revolution of all the particles is independent of their distance, and depends on nothing but their density or closeness together; and the time varies inversely, or the velocity of revolution varies directly, as the square root of that density. And it may easily be proved that this holds not merely for the particles revolving in great circles, or in a plane through the centre of the sphere, but also for those revolving in small circles like parallels of latitude round a polar axis.

So we have this remarkable result, that a sphere of matter equally dense throughout, whether its particles are free as in a fluid globe, or freer still as in a nebulous one, can all revolve together, with equilibrium between the centrifugal and attractive forces, without any part wanting to lag behind the rest as they would in a flat ring like Saturn's (p. 190). Then if the inner parts alone contract, and become denser than the outer, they will revolve faster, for the reason given at p. 234 ; but the outer parts will revolve at the same rate as before ; for the mass of the sphere within remains the same, and attracts as if it were all condensed into its centre, though the density is no longer uniform. Consequently the outside of a nebulous globe densest in the middle must revolve slower than the inner parts, if equilibrium is to be preserved. Also the particles of a revolving nebula which is contracting must evidently move in spirals toward the centre, and get denser as they get nearer; which is the aspect of the spiral nebulæ. But mere rotation round a polar axis would not prevent the globe from collapsing into a flat spheroid by the attraction of its particles toward the equator. They can only be kept apart by the expansive force, or else by revolving in great circles in all directions.

Another consequence of the law of time and distance and mass at p. 251 is that a dense globe of stars or loose matter must turn faster than a thin one of the same size to preserve equilibrium. This may seem strange, but you must not judge of bodies which contain their own force by anything we can deal with on the earth. We saw that the time varies inversely as the square root of the mass, and therefore of the density of a mass of given size; and therefore the velocity of rotation varies directly as that same square root. So if the earth's orbit were nearly filled up with a sun of that size, and of the same density as the present sun, the earth would have to go round it in 2h. 45m. instead of a year, to avoid falling upon it and being stuck there. And a globe of little planets or any other loose matter of that average density, filling up the orbit of the earth, must rotate with that velocity. But on the other hand, if the sun were spread out into a sphere large enough to fill the earth's orbit, it would take a year to rotate (p. 270). None of the spiral nebulæ turn fast enough for their motion to have been perceived yet, though some double stars revolve in 30 years: which is therefore a proof of the extreme thinness of these nebulæ, if they do rotate. Of course a solid or a fluid globe, whose internal attraction can compress it no more, may turn as slowly as it likes, though not too fast, or it will break, and throw off its equator. The earth might turn 17 times faster than it does before it would throw off its equator, which now only swells out 5 miles beyond the radius of an equal sphere (pp. 24, 37).

THE STARS.

I mentioned the parallax of the stars just now; and I explained at p. 208 that the parallax of the sun, or moon, or any planet, is the apparent radius of the earth as it would be seen from those places; and that it is found by observing the difference of apparent position of those bodies from opposite ends of the earth's diameter. But it is of no use to look at the stars from opposite sides of the earth for the purpose of measuring their distance. A very few of them can just be perceived to change their places with reference to the rest, when looked at from opposite ends, not of the earth, but of the earth's orbit, or at intervals of half a year. That therefore is our scale for measuring their distances; and their parallax is the apparent radius of our orbit, as it would look to them, or the angle whose numerical value is our sun's distance divided by their distance. There is only one star, *a* Centauri, so near that our orbit would appear to it as wide as the planet Neptune does to us, its parallax being 0″·976; the next, 61 Cygni, is nearly twice as far: two stars called 21,238 Lalande and 1830 Groombridge, 3½ and 4 times; 70 Ophiuchi, *a* Lyræ, and Sirius, all above 6 times as far; Arcturus and two or three more have a just sensible parallax.

Now let us see what this means in actual distance. The numerical value of 0″·976, by the method I explained at p. 241, is ·0000047; or the nearest star is 211,340 times as far off as the sun, or about 19½ trillion miles. And as light takes 8¼ minutes to come here from the sun, you may easily calculate that it takes 3⅓ years to come from the nearest star, 6½ years from the next nearest (as far as we know), and 21 from Sirius the brightest star; and of course much longer from the fainter stars, which there is good reason to believe are on the average more distant than the bright ones, as we shall see presently.

Magnitudes of the Stars.—The stars are proverbially 'innumerable,' like the sands on the sea shore. Even the number visible with telescopes is beyond all counting. It is estimated roughly, by dividing them into districts and counting a few, that not less than 5 millions can be seen with a telescope far below the largest; 24 of the brightest are called 'stars of the first magnitude,' 60 of the second, 200 of the third, and about 15,000 of the smaller magnitudes above the eighth. These divisions are merely arbitrary; but now that their brilliancy can be compared by experiments, Sir J. Herschel and other astronomers have proposed that some more definite scale of magnitude shall be agreed on. But their brightness or apparent magnitude bears no definite relation to their distance, or even to the intrinsic brightness of the few stars whose distance has been measured. Sirius is 4 times as bright as *a* Centauri, though 6½ times as far off; therefore it must be 170 times (or $4 \times 6\cdot5^2$) as bright at the same distance; and Sirius gives out nearly 400 times as much light as the sun. It has been estimated that the light of sunshine here is a hundred million times as much as we receive from all the stars.

The principle which Sir J. Herschel advocates for classifying the stars by magnitudes, *i.e.*, by apparent brightness, and to which the received classification does approximate, is that stars of the second magnitude should be those which are on the average twice as far off as those of the first, as far as we can judge; third magnitude stars three times as far as the first, and so on. But how is that to be done? It can only be done on the assumption that the stars generally give out the same absolute quantity of light, or would all look equally bright if they were at the same distance; which means that the intrinsic or specific brightness × the real size of the disc of each star is the same; which it certainly is not individually, as we saw just now, but may be nearly enough for a general average. We must also assume for this purpose what is rather improbable, that no star light is worn out or lost in its passage through the luminiferous æther, as sound is by the air which carries it, especially when there is any wind against it: a fact which no philosopher has yet attempted to explain, as far as I know. Bells which are heard 6 miles off with the wind cannot be heard a mile against a wind which you can hardly feel.

Assuming these two things, the light of the stars here will vary inversely as the square of their distance; and therefore a star of the second ' magnitude ' must be one whose apparent light is only a quarter of those of the first magnitude; the light of third magnitude stars is one ninth of the first, and so on.   The proportions of two lights can be pretty accurately measured by contrivances for cutting off so much of the brighter one as will reduce it to an equality with the dimmer one; which the eye can judge of, though it cannot compare proportions of light without such aid.   In this way then the distances of the stars visible by the eye can be approximately measured in terms of the distance of the nearest, *a* Centauri, which is measured by its parallax of nearly 1″, or as multiples of 20 trillion miles.

But we can go a great deal farther than this with our measures.   An eye pupil twice as wide as usual would take in four times as many rays of light from the same star at a given distance, or would take in as many rays from a star at distance 2 as a common eye at distance 1, since they spread out four times as wide at twice the distance.   You will see in the next chapter that telescopes only penetrate into space by virtue of their being a very large eye, or bringing into our eye much more light from a star than it can collect for itself.   Therefore if a telescope which penetrates into space 200 times farther than we can see without it, brings into view a new star, making it as bright (say) as one of the 8th magnitude, it must be 200 times as far off as the average of 8th magnitude stars, upon the assumptions we have made, or 1600 times the 20 trillion miles which is the unit for star magnitudes or distances. And as the light takes $3\frac{1}{3}$ years to come here from the nearest stars, it would take 5300 years from such a star as I have described, of which there are multitudes beyond counting, and others far beyond them.

' As the stars recede into distance, time recedes with them ; and there are stars from which Noah might be seen stepping into the ark, and Eve listening to the temptation of the serpent.'* How far they do recede we cannot tell : all we know is that no telescope has yet reached the outside of them ; for every increase of telescopic power only brings more stars into view.

If light is wasted or worn out in its passage through space, as it is through a fog, these calculations fail to this extent, that the light decreases faster than the square of the distance increases, and we do not know how much faster. In that case stars beyond a certain distance can never be seen in any number of years, because the undulations of their light would be all gone before they reached our distance. Indeed by the ordinary law of emanations according to the inverse square of the distance, the rays from any single star may become too thinly spread to make any distinct impression on the eye, even when gathered in by any possible telescope ; but a multitude of stars then combine their light and produce a luminous haze or nebula, which may or may not be resolvable by telescopes.

If all the stars above the horizon at night, visible and invisible together, cover an apparent space equal to the sun, and are intrinsically as bright, their light does waste prodigiously in coming. Otherwise the

* Froude's Lecture on The Science of History.

night would be as bright as the day, for the reason given at p. 165. And if the sky is so full of stars that we nowhere see right through them, the earth would get 455,600 times the light and heat it does from the sun, unless they waste in coming.* The question whether light is lost depends on whether the luminiferous æther is perfectly elastic, *i.e.*, whether its particles move among themselves without friction or viscidity : a wave will go farther in water (though even that has some friction) than in treacle.

As a globe as wide as the earth's orbit would have a scarcely measurable diameter to any of the stars, it is not to be expected that any single star should have a measurable size to us. Every now and then people have fancied that they could discern an apparent disc of some very bright star, but it has always turned out to be what is called the *spurious disc* with rings round it ; which is one of the consequences of the undulatory theory of light, and is larger with a small object glass than a large one. A planet looks larger in a telescope, because even a disc of 2″, like Neptune's, can be magnified into an angle 1000 or 2000 times greater ; but a telescope only shears a star of the rays which give it a false appearance of size to the eye. The twinkling of the stars also cannot be explained without a fuller account of the undulatory theory of light than can be given here. Again, when the dark edge of the moon comes over a star it is extinguished suddenly ; which not only proves that the moon has little if any atmosphere, but that all the rays from the star to our

* Appendix, Note XXVI.

eye are practically close together, as if it were a point, even at the distance of the moon.

Many stars are not single, but consist of two or three so close together that they require a high magnifying power to separate them, though their distance apart must be enormous to show even then ; and they revolve round each other. About 600 brilliant double stars are known, besides the more numerous fainter ones ; and in 120 of these pairs the companions have different colors, besides different degrees of brightness. And both the brightness and the color sometimes change during their revolution, as if their opposite sides were different. Many single stars change also. Sirius (which however is double) was called a red star 1700 years ago, in the time of the astronomer Ptolemy, the author of the erroneous theory of the solar system which is called after him ; but it is now a white one. Several others change periodically, and a few rotate like the sun. In 1866 a star in Corona Borealis rose quickly from the 9th to the 2d magnitude, and soon after faded again, as others have done before. Everybody has heard of the 'lost Pleiad,' by which those once conspicuous ' seven stars ' are reduced to six. And many others have come in and gone out within the time of astronomical history.

Spectrum Analysis of Stars.—The star which started into temporary splendor in May, 1866, and again from August till November, was analyzed by Messrs. Huggins and Miller by the method I have several times mentioned, and its spectrum showed a great outburst of hydrogen gas. A good many other stars have been

examined in the same way, and are found to contain
sodium (or the base of common salt), magnesium, mercury, antimony, bismuth, hydrogen, and some other
less common materials, including a few not known
upon the earth.   Their light, like the sun's, comes from
some solid incandescent matter surrounded with an atmosphere containing the vapors of these elements. The
same investigators say that blue or purple stars are
always found in connection with a stronger red one;
and there is no doubt that the difference of color in the
stars is due to chemical differences of composition, and
temporary changes of color always show a change in
the condition of their atmosphere.

Double Stars.—The question whether the second and
paler of a pair of stars is a planet or a star, means
simply, is it self-luminous, or does it only shine by reflection ?   For in all other respects there is no difference, although we are accustomed to think of planets
as much smaller companions of a sun than the companions of some of the stars appear to be.   If any star
is found to have a satellite or companion which is periodically brighter and dimmer, and quite invisible at
that part of its orbit which we call inferior conjunction
for Venus and Mercury, when they are between the
sun and us, that would be a planet.   The author of
the ' Plurality of Worlds' remarks that the periodical
obscuration of the star Algol does not prove the passage of a planet over it; for the planet would have to
be one-sixth as wide as its own orbit to produce the
observed effect; which is altogether impossible.   But
it seems now tolerably certain that Sirius has a compan-

ion or satellite: which may be called a planet from its darkness, but is very unlike one in its mass; for its existence was discovered by its causing a large *nutation* backward and forward in the mean motion of Sirius, which I shall speak of presently.; and the planet has been since observed, and its distance measured.

In this case, and in all the revolving pairs of stars, whenever they are within measurable distance, and there has been time to see the period of their revolution, we can also ascertain their joint masses, and sometimes even their separate ones. For by the law of time and distance, of which I have said so much, the period of two bodies revolving round each other depends only on their mean distance and their joint mass. So if their distance can be got, clear of foreshortening from an oblique position of their orbit, that and their period will give their joint mass. And then the proportion of the masses to each other can be found in the way described at p. 251 for the earth and moon, provided we can measure also how much either of the two stars oscillates about the mean place or common centre of gravity.

Let us illustrate this by performing the operation on Sirius, assuming that the facts stated are correct. His companion is said to be at least 47 times as far off as we are from the sun, or above half as far again as Neptune, and the period is about 49 of our years. Then if you look back at p. 250 you will see at once that $\dfrac{\text{Sirius}}{\text{sun}}$ must $= \dfrac{47^3}{49^2} = 43\cdot25$; remembering that 'Sirius' here means the mass of himself and his com-

panion together, as 'sun' means really sun + earth. But astronomers have also made out that Sirius oscillates, or nutates, on each side of the joint c. g. $16\frac{1}{4}$ times our mean distance; and therefore the joint mass has to be divided between them in the proportion of nearly 31 to 16; and we may say that Sirius alone =
$$\frac{31 \times 43 \times \text{sun}}{47} = 28 \text{ suns.}$$
This, however, is very far short of the measure given by their relative quantities of light; for that proportion of mass (if the densities are equal) gives a diameter only 3 times, and a disc 9 times the sun's, instead of 400, which is found to be the proportion of their lights; but Sirius may be made of materials lighter (in both senses) than the sun.

Similarly the elements of three other double stars have been found, and we may arrange them as follows, making the sun the unit of mass, and our year and mean distance the units of period and distance between the two associates. I have not the data (if they exist) for separating the masses of any but Sirius, as we did just now.

	Joint mass.	Period.	Distance.	Parallax.
Sirius	43·25	49	47	″·15
70 Ophiuchi	3·1	96	30·5	·16
a Centauri	·55	78	15	·98
61 Cygni	·1	514	29·3	·54

Motion of the Stars.—There remains the question, are the stars generally in motion; and if so, what kind of motion? That can be answered *à priori* thus:— Nothing can be at rest where attraction is universal. Not rest but motion is essential to the stability of the universe; or rather we must say, it is a choice of motions; either the stars are all moving to destruction and chaos at a common centre, or else they keep their distance only by revolving round it. Whether they extend an infinite or a finite distance into space, which must needs be infinite, as time must run into eternity, can make no difference; for each successive shell of stars does nothing to help those within it against their mutual attraction to the common centre, since a spherical shell exerts no attraction within it (p. 29). And so the whole starry sphere, whether we like to consider it a large or a small one, must revolve under the conditions I explained before for any single nebula or globe of stars. That assumed indeed that they are scattered uniformly through the sphere : which they are not, for some parts of the sky are much more starry than others, especially the Milky Way; but that only varies the forces, and affects the direction and velocity of the motions, and not the argument.

But the question has been answered by observation also. Sir W. Herschel discovered that one important star, the sun, moves toward the constellation Hercules, at a rate which is now estimated as about the diameter of the earth's orbit in a year. And lately a general motion of them all, but one, is thought to have been ascertained. That one is Alcyone or η Tauri of the Pleiades (p. 52), which is therefore assumed to be the centre of the universe. But Sir J. Herschel thinks that unlikely, because Alcyone is 26° away from the Milky Way, which can hardly be conceived according to dynamical laws to revolve in any direction but in its own plane, as a grindstone does, and therefore not round a centre so far off.

Not that the stars of the Milky Way need revolve in the same time like the particles of a grindstone, or are likely to revolve round any polar axis. For a mass of stars so revolving would have sunk into the plane of the equator through the common centre of gravity long ago by their mutual attraction, unless indeed they keep vibrating across it. Each star probably moves in an independent plane passing through the common c. g., as the planets move in planes through the sun all within the narrow band of the zodiac. Indeed if the solar system and the other stars did revolve in the same time round any polar axis, we could no more see that any of them moved at all than one part of the earth can see the rotation of the other parts. But the theory of the motion of the stars is in a very imperfect state as yet.

It is worth while to calculate the kind of period which a revolving mass of stars would require for equilibrium between attraction and centrifugal force, on some simple assumption of the average density of the mass. Let us assume then that all the stars are as heavy as the sun on the average, and generally as near each other as the nearest of them is to us, *i.e.*, 20 trillion miles off: which we will write 2 ⑬ for shortness, meaning thereby 2 with 13 cyphers, which only have to be doubled and trebled for squares and cubes. This assumption is probably too favorable to the average density of the universe, but it will answer well enough for the purpose of illustration. But how close can the stars be packed, or can they be packed at all, so that all the intervals may be the same? The first of these questions is not so easy as it looks, and I can only give you an answer which is the result of some trigonometry.

If you want to see how the stars must be arranged, to be at equal distances, get some bullets and lay three together in a triangle, or four in a rhombus (not a square), and go on from that making as large a heap as you like. In this way each star stands in the middle of a sphere of 12 others equidistant from it and from each other. If any number of balls are so packed in a rhomboidal box which exactly fits them, it may be proved that each ball occupies on the average ·707 of the cube of its own diameter; while the ball itself is only ·523 of that cube, as I said at p. 15. Therefore in order to find the average density of the universe, we must consider the matter of each star to be expanded into the bulk of ·707 × 8 ⊙ cubic miles.

But we saw at p. 270 that the sun would rotate in a year if his present matter filled the earth's orbit, or were expanded into a globe of ·523 × 914³ ⊕ cubic miles; and that the time of rotation of nebulous globes is inversely as the square root of their densities. Consequently the time of rotation of a globe of the average density of the universe is to a year as the square root of the larger of the above masses is to that of the smaller; which you will find, if you take the trouble to work it out, is a little more than 119 millions to 1. Therefore that is the number of years which the whole mass of stars would take to revolve, on the assumption

13

we have made : and still more if the average density is less. And as 360° = 1,296,000″, this amounts to a revolution of 1″ in nearly 92 years.

The central attraction on a particle of a very flat spheroid of uniform density also varies as its distance from the centre, as in a sphere; though the amount of attraction is of course less than at the same distance from the centre of the sphere, because the attracting mass is less in the proportion of the minor axis to the major. But the attraction is 2·2 as much as it would be according to that proportion. Therefore if the Milky Way is equivalent to a spheroidal mass of the density we assumed, and with a major axis = 10 times the minor, its internal attraction will be ·22, and the velocity of revolution √·22 or about half of what we calculated for the sphere.

THE CELESTIAL GLOBE.

The thing called a celestial globe is a picture on a globe of all the stars as they appear in the sky. Properly of course such a globe should be hollow, and we in the inside of it. And in fact all the pictures *are* reversed from what they appear in the sky. Take the well-known figure of the Great Bear or Charles's Wain, which is much more like the outline of a saucepan with a bent handle. As we see it in the sky, and in star maps, this handle, or the bear's tail of three stars, is on the left hand (except at that time of night when it is turned wrong side up) and the two stars which form the outer edge of the saucepan, and are the pointers to the north pole star, are on the right hand. But it is drawn the opposite way on the outside or convex surface of the celestial globe. The celestial globe is covered all over with pictures of those Bears, Lions, Virgins, Scorpions, and other things, into which the ancients fancifully divided the groups of stars and called them constellations. The particular stars are usually denoted by Greek letters attached to the Latin name of the constellation, as a Alpha Lyræ (the brightest northern star), a Centauri (the nearest), β Beta Leonis, γ Gamma Draconis, and some by numbers in a constellation, as 61 Cygni, etc. A few have names of their own, as Sirius, Alcyone, Polaris, now the pole star within 1° 24', but once 46° off in consequence of precession ; and it will be as near as 26' in 360 years.

Another thing which a celestial globe has on it is *the Zodiac*, which is a band of 8° on each side of the ecliptic, or 16° altogether, and on which are constellations of the 12 Signs of the Zodiac, and within which all the planets have their orbits, except some of the Asteroids. But as I explained in speaking of the precession of the equinoxes (p. 51), the signs have left their constellations nearly 30° within the time of astronomical records. Therefore you see on the celestial globe the two crossings of the equator and ecliptic, ♈ and ♎, lying toward the left hand of the picture of the Fishes and of the lady called Virgo, the Astræa of the poets, who holds the scales of justice (Libra) in her hand. Celestial globes, like terrestrial, are made to turn upon the poles of the equator, not on the poles of the ecliptic. One would be just as right as the other for a celestial globe, only they represent different things ;

turning the globe on the poles of the equator represents
the daily rotation, which produces the rising and setting of the sun, moon, and stars : turning it on the
poles of the ecliptic would represent the motion of the
moon round us in 29½ days, and the apparent motion
of the sun round us in a year.

Right Ascension and Declination.—I have already said
at p. 48, that longitude of the heavenly bodies is measured from ♈ on the ecliptic, and not on the equator
from the meridian of Greenwich : which indeed would
be nonsense, as Greenwich is changing its position
with reference to every star every minute. Perhaps it
might as well have been measured on the equator, as
another name has had to be invented for the measuring of the stars along the equator from that point ♈
where all the celestial measures of that kind begin.
That other name is *Right Ascension*, commonly written R. A. ; and there is no east and west in it : it runs
on from 0 to 360° on the celestial globe, and from left
to right, to correspond to the earth's moving as it does
through space from right to left. The thing which
corresponds to terrestrial latitude for the stars is called
declination : that is their distance from the equator.
R. A. is also measured by time, like terrestrial longitude, 15° 15′ 15″ being = 1h. 1m. 1s.

There is also *heliocentric* longitude, which is the angle between two planes through the sun's centre and
perpendicular to the ecliptic, one through the planet
and the other through ♈ ; while *geocentric* or common
celestial longitude is the angle between two such planes
through the earth's centre. So the terrestrial longi-

tude of any place is the angle between the two planes called meridians, which both go through the earth's centre, intersecting all along the polar axis, one through the place in question and the other through Greenwich observatory, or whatever place each nation takes for its zero or 0° of longitude. The *declination great circles* for R. A. also intersect in the polar axis, only the zero is at ♈ ; but the *quasi-meridians* for celestial longitude intersect in the poles of the ecliptic. Perhaps I should remind you that the angle between (or the inclination of) two planes contains the same number of degrees, minutes, and seconds as the arc which the planes cut out of any circle of any radius drawn round their line of intersection as an axis, such as small circles of latitude no less than the equator. Of course the larger you make the circle the larger the arc is—in magnitude, but not in degrees, as the minutes on a clock face are larger than on a watch.

Polar Distance is the *complement* of declination, or the difference between the declination and 90°. This is so much used that it is abbreviated in almanacs for our hemisphere into N.P.D. So the *zenith distance* of a star is the complement of its *altitude* above the horizon. The distance of the pole, or pole star (if it were exactly there) above the horizon, or the distance of the zenith from the equator, is the latitude of the place where you are; and the distance of the pole from the zenith is the *colatitude,* or the difference of the latitude from 90°.

The 'horizon' in astronomy does not mean the accidental boundary of your sight by earth or sea, but a plane through the middle of the earth parallel to the 'level' surface of water or mercury where you are, or to which the plumb line is perpendicular (see p. 12). The *dip of the horizon* is the angular distance that you can see below the hemisphere of sky of which the base is the level plane through your eye; and it is evidently greater the higher you are above the earth. The lowest point, opposite to the zenith, is called *nadir ;* both of which words came from Arabia, like the whole contrivance of our common numbers, without which the science of arithmetic and mathematics must almost have stood still.

The word *Azimuth* also came from there; which is to Altitude what Longitude is to Latitude, and what R. A. is to Declination; viz., the distance between two great circles like meridians, passing through the zenith instead of the pole; and it is reckoned from the north point of the horizon; but the *Prime Vertical* goes through the zenith and the east and west points. *Celestial latitude* is measured from the ecliptic, as terrestrial latitude is from the equator. Formerly the longitude of the planets was, and in some almanacs still is, given by the signs, as ♎ 25°, or ♓ 6°, but more frequently now by the corresponding number of degrees, 205°, or 336°. Ancient dates are sometimes fixed by certain stars having been said to rise *heliacally* then : which means rising just before the sun.

An ecliptic is generally drawn on terrestrial globes, crossing the equator at 30° west and 150° east longitude: but it has no meaning there. On the celestial globe it has a meaning and a proper place; and the

great circle half way between ♈ and ♎, which passes
through the poles of the equator and ecliptic and the
places of the solstices, is called the *solstitial colure.*

Orreries—so named after a Lord Orrery who bought
a *planetarium* 150 years ago—are of great antiquity,
and may be called working models of the solar system.
In old times, of course, they made the earth its centre;
but they were made by Huyghens and Romer to illustrate the Copernican system and the motions of Jupiter's moons, and much later by Dr. Young and other
astronomers. By a proper combination of wheels, of
which I gave a simple illustration at p. 57, they can
be made to show all the regular motions of the earth,
moons, and planets, the changes of seasons, and the
conditions of eclipses. But an orrery with a moon
only a quarter of an inch wide and an earth of an inch
ought to have a Jupiter of 11 inches, a Saturn's ring
of 21, a sun of 9 feet, and a platform of 11 miles diameter to take in the orbits of all the planets on their
proper scale.

CHAPTER VI.

Books on astronomy generally give some account of the modes of mounting and using telescopes for different purposes, while the explanation of the principles or 'theory' of the telescope itself is left to treatises on optics.* But for that simple looking tube, with only a large magnifying glass at one end and a small one at the other, we should hardly know more of astronomy than the Chaldæans did, or at any rate Copernicus, who only guessed but could not prove that the sun is the centre of the solar system; nor should we be able to measure a single celestial distance, or even the earth itself accurately, or navigate the great seas with certainty of reaching any given place, and the civilization of the world would stand still. Therefore I propose to give a fuller explanation of the theory of telescopes than in the former editions of this book, so far as it can be given without resorting to mathematics.

I said at p. 35 that Galileo was the first inventor of telescopes; and so he was for astronomical purposes, though it appears that they had been made above three centuries before in England by the famous Friar Roger Bacon, the greatest philosopher from his own time to

---

* There is a clear and concise treatise on optics in Chambers' Educational Course, requiring very little knowledge of geometry.

that of his more celebrated namesake Francis Bacon, Lord Verulam and St. Albans, commonly called Lord Bacon, which he never was. Friar Bacon also invented gunpowder, and so many other things that he was accused of magic, like Dr. Faustus, one of the inventors of printing, and imprisoned by the Pope for ten years —much longer than Galileo, and only released when he was 74, after which he lived six years, and died at Oxford in 1294. But the invention perished with him, and was lost till it was made again by Jansen and Lippershey in Holland, in what form is not known, and very soon after independently by Galileo in 1609, in the form which still survives in opera glasses.

· That was however soon superseded by what is called the ' astronomical telescope,' as the best kind of watches are called chronometers, though all watches are or profess to be chronometers or time-measurers. Kepler seems to have suggested this construction first, though he did not execute it, nor perceive all its advantages. One Father Scheiner was the first who did, and it was afterward improved by Father Rheita ; and still more in 1656 by the celebrated Huyghens, who discovered the true law of refraction, and also invented a compound eye glass which is still for some purposes the best. For the present I only say that Galileo's telescope is a short tube with a large convex lens, or one thickest in the middle, at one end, called the object glass, and a small concave one called the eye glass at the other end. The ' astronomical' differs from it in having a convex eye glass at the end of a longer tube. In both, the object glass must be flatter than the eye glass to produce any

13*

magnification : indeed the amount of it depends entirely on the proportion of the curvature of the eye glass to that of the object glass.

Space-penetrating Power.—But before we talk of magnifying there is another part of the business of telescopes, which is much more simple, though less understood by most people, who are satisfied with knowing that a telescope is a combination of glasses for making distant things look nearer by magnifying them. It is not by means of magnification that telescopes enable us to see millions of stars beyond what the eye alone can see : or as it is called, to penetrate farther into space. The *space-penetrating power* depends simply on a telescope being a very large eye, and bringing into our eye as many more of the rays from each point of a distant object as the area of the object glass exceeds the area of the pupil of the eye. It is true that a telescope cannot bring into the eye all the rays it receives unless the magnifying power is at least equal to the width of the object glass divided by the width of the pupil ; but the magnification of powerful telescopes far exceeds that.

To measure that power we must remember that we should want an eye four times as *large*, or twice as *wide*, to see a star just visible now, if it were removed twice as far ; since the rays are spread out four times as thin at twice the distance. Besides that, at least one-eighth of the light is lost in passing through the glasses ; which is the same as if the area of the object glass were only $\frac{7}{8}$, or its width $\sqrt{\frac{7}{8}}$ or ·935 of what it really is. Therefore the space-penetrating power of a tele-

scope is to that of one eye as ·9 of the width of the
object glass is to the width of the pupil—nearly a
quarter of an inch.

By far the largest telescopes that are made have
concave mirrors instead of object glasses, as I shall explain afterward; and their space-penetrating power
would be greater in the same proportion as their width;
but much more light is wasted by reflection than by
passing through glasses : indeed nearly half generally
instead of one-eighth. Therefore the space-penetration
of Lord Rosse's telescope, whose mirror or speculum is
6 feet wide, is $\dfrac{1}{\sqrt{2}} \times 72 \times \dfrac{4}{\sqrt{2}}$ or 144 times that of a
pair of eyes ; for the power of two eyes is $\sqrt{2}$ times that
of one,* on the same principle as just now stated. This
means that the telescope will penetrate a sphere of
stars of 144 times the radius that we can see without
it; and therefore, if the stars are scattered equally
thick all through what we can only call infinite space,
we can see 144^3 or nearly 3 million times the usual
number with such a telescope, except so far as they hide
each other. The penetration of a refracting telescope,
i.e., one with an object glass instead of a mirror, 15
inches wide, the largest yet completed, is about 47
times that of a pair of eyes by the same calculation.

Magnification.—In both cases the whole of the rays
received by the telescope are made to converge by the
mirror or object glass in just the same way as the rays
from the sun converge into a small bright spot or image
of the sun in a common burning glass. But this is con-

---

* Appendix, Note **XXVII.**

nected with the other part of the business of telescopes, which requires a good deal more explaining. Magnification, either by a single convex lens or by any combination of glasses, consists in the rays which come to the eye from the outsides of an object being made to come at a wider angle with each other than they do naturally. But that is not all : if it were, a common 2 inch magnifying or burning glass which makes a bright spot at 3 inches from the glass would magnify the sun from 32′, its apparent diameter to the naked eye, to 37°, or nearly 70 times: that being the angle at which the outside rays converge to a focus through such a glass.

Indeed if you could look at the sun through such a glass with your eye at the focus—and you may at the moon—it would appear magnified enormously; but then it is all in confusion, because the rays from each point of the sun or moon are not brought to a point again at the back of the eye, where a distinct (though inverted) picture of everything seen must be formed if it is to be seen distinctly. The picture of the last thing seen actually remains there for a time after death, as Scheiner discovered long ago. The condition for distinct vision is that the bundle or *pencil* of rays which come from *each* point of the object to the glass shall be brought to the eye in a parallel state, like a thin stick of straight wires (or rather diverging for short-sighted eyes), but the pencils from the *various* points of the object must make some sensible angles with each other if it is to have any apparent size ; and the wider those angles are the more it is magnified.

It may occur to you to ask what is the use of apply·
ing telescopes to the stars when no magnifying power
will make them look any larger (p. 283). The use is
to measure their distances from each other and from
the meridian or the equator or any other circle; for
those of course are magnified by the telescope, and are
measured by the time the star takes to move over the
distance, or by the angle which the telescope has to be
moved through from one star to another.

Again you may say that terrestrial objects seen
through a telescope do not appear larger than usual,
but only nearer. But that is exactly the same thing.;
for, as I have said several times, the apparent size of
anything has no meaning except with reference to its
distance; and it is only because we know by experience the size of houses, trees, and men, that we do not
think of them as appearing larger when seen near than
far off. When we come to mountains or buildings
above the usual size we are constantly deceived, and
generally take them to be smaller and nearer than they
are. On the other hand very small things do appear
very large through a microscope, which is only a telescope adapted for near objects, because we can never
see them so large with the naked eye, in consequence
of the eye not being able to see anything distinctly
nearer than 5 or 6 inches; for then the rays from each
point diverge too much to be brought to points again
by the lenses of the eye, and to form a distinct picture
on the retina.

The short explanation of an astronomical telescope
is this: see the figure at p. 312. Rays of light spread

from the point A of the object ♀ in all directions. A certain number or *pencil* of them is received by the object glass OPQ all over it: after passing through that convex lens they converge to a point or focus *f*. In like manner the rays from another point B come to the object glass, crossing the A rays without at all interfering with them, and are made to converge to another point *i* at the same distance as *f*; and in that way an inverted image *fi* of the object AB is formed there, which is as much smaller than the object as O*f* is less than OA, and very near the eye glass *abc*. Then the rays that came from A diverge a little from *f* to *a*, and the B rays to *b*, and there both pencils are refracted and made to converge into the eye at E, making a much wider angle *aEb* with each other than they did coming from the object: which is the increase of apparent size or magnification. Besides that, the rays in each pencil are changed by the eye glass from diverging into parallel ones, which is the state the eye requires for distinct vision.

Law of Refraction.—But though this is enough as a general description of a telescope, I have given you no reasons for anything that the lenses do, nor explained what the magnifying power is for any given pair of glasses. If you want to understand these things we must begin again with refraction, and see how they all come from one tolerably simple law.—not to go farther back into the causes of that, which are far more complicated. For this purpose we must use a figure. A BC is a prism or wedge-shaped piece of glass, and S*a* a ray of light entering it at *a* from the air, and X*a*Y perpendicular to the surface. Then instead of the ray going on straight, it is bent aside at *a* into the direction *ab* nearer the perpendicular, unless it happens to be itself perpendicular, in which case only it goes straight.

So long as the obliquity is small, as it always is with the rays which fall on telescope lenses, the angle S*a*X in the air may be considered to bear the constant proportion of $\frac{3}{2}$ to the angle *ba*Y in the glass; which number is therefore called the *index of refraction* for glass; as $\frac{4}{3}$ is for water (compared with air) and 2·44 for diamonds. At *b* the ray emerges again out of glass into air, and follows the same law, turning off into an angle G*b*Z which is $\frac{3}{2}$ of the angle *ab*Y in the glass, Y*b*Z being the perpendicular at that surface. The angle in the thinner medium is generally called the 'angle of incidence' and that in the dense one the 'angle of refraction;' but they are just the same whichever way the ray is going, and it is better to think of them as the air angle and the glass angle.

If the two surfaces of glass were parallel, it is evident that the second refraction at *b* would just balance the first at *a*, and the ray would emerge parallel to its first direction, only a little pushed aside, according to the thickness of the glass and the obliquity of the in-

cidence on either surface. But if the surfaces are inclined, the emerging ray (whether it is S*a* or G*b*) devi-
ates from its original direction, as you see in the figure,
and is brought downward toward the thickest part of
the glass, or away from the angle C, called the *refracting angle*, where the two refracting surfaces meet. It
may be proved by a little geometry that the deviation
is half the refracting angle of the prism ABC; *i.e.*, if
S*a* is parallel to BAG or any other line in that sort of
position, the angle *b*GA will be half the angle C, provided the angles at *a* and *b* are small ones; and they cannot both be small unless C is.

Though this is not a treatise on optics, yet to prevent
mistakes I had better tell you that when the angles of
incidence and refraction are not small the relation between them is not quite so simple. If you draw any
line SX at right angles to *a*X, the fraction $\dfrac{SX}{Sa}$ is
called the *sine* of the angle S*a*X. (In old books SX
would be called ' the sine of S*a*X to radius S*a*,' but the
other is the most convenient definition and the least
liable to mistake.) And the true relation between the
two angles, either at *a* or *b*, is that the sine of the air
angle is $\frac{3}{2}$ of the sine of the glass angle. You will find
in books of mathematical tables the sines of all angles
up to 90°, when the $\dfrac{SX}{Sa}$ evidently becomes 1, and they
can go no farther, but decrease again; but so long as
the angles are small, the sines increase in very nearly
the same proportions as the angles themselves, which
simplifies telescope calculations materially.

Internal Reflection.—On the other hand when you get up to 41° 48′, the sine of that is ⅔; and therefore no glass angle beyond that can have any corresponding air angle, as there is no angle with a sine greater than 1. And then comes the remarkable consequence that a ray trying to emerge from glass at any greater obliquity than that, which is called the *critical angle*, cannot get out there at all, but is reflected back again, more completely than if the back of the glass were silvered for a mirror. Thus if the angle abY in the figure were as much as 42°, the ray would not go out of the glass, but would be reflected to c on the third side of the prism, and there go out to E with the usual refraction, unless it arrived at c also too obliquely to get out, when it must try again at the side BC. The critical angle of water is 48½°, and of diamonds only 28° : which is the reason of the brilliancy of diamonds, and of dewdrops which stay on certain leaves unbroken, viz., that a great deal of the light falls on them too obliquely to pass through their back surface, and so it is reflected.

Reflecting Prism.—Therefore if light falls directly (*i.e.*, perpendicularly) on one of the narrow faces of a prism whose angles are 90°, 45° and 45°, it reaches the wide face at the same angle 45°; and as that exceeds 42°, it is all reflected out directly at the other narrow face, at right angles to its original direction ; and such a prism is a far better reflector than a flat mirror set at the angle 45° to reflect at right angles. You will see at p. 332 the use which is now made of this in Newton's telescope. A little light indeed is reflected at every change of surface, which is the reason of the loss of light in passing through the glasses of telescopes, but only a little until the critical angle is reached. You will see how this also is turned to account in telescopes for looking at the sun (p. 333).

If you find it difficult to believe that light cannot pass in any direction through a transparent medium, you may verify it by hanging something a little under water on the far side of a trough or basin, and looking at it while you bring your eye down to the water on the near 'side. At a certain distance it will vanish, although you can see the far side of the trough below and beyond it. The reason is that the rays which ought to come from it to your eye reach the surface of the water too obliquely to come out, and are reflected down again internally, as you would see if your head were under the water instead of over it.

Thus far there is no material difference between a thin prism of which ABC is a section and a convex lens; for though its faces are segments of spheres (either alike or not), a ray passes through it exactly as it would through a prism touching it at the points *a*, *b*, as I have drawn the figure. But when many rays fall on different parts of the surface, a lens will manifestly affect them all differently; and those which fall near the edge will be more refracted than those near the middle, because the surfaces are more inclined to each other near the edge than near the middle, where they are parallel. And it fortunately happens that all the rays from any one point, far or near, after passing through a convex lens, whether directly or obliquely, converge again (very nearly) to another point or focus somewhere in a straight line through the first point and the centre of the lens; which line is called the axis of that pencil: as you may see with the rays spreading from A or B over the lens OPQ in the astronomical telescope at p. 312, and converging again to *f* and *i* in the axis of each pencil.

Focal Length.—If the rays come from a point so far off that they are practically parallel, the point of convergence is called *the* focus of the lens, and its distance from the lens is called the focal length. Conversely if rays emanate from a point at the focal length, they will emerge from the lens parallel; but if from a point nearer, they will not converge at all, but only diverge less than before: if from a point beyond the focal length, they converge to a focus somewhere beyond the focal length on the other side. These two points of divergence and convergence are called *conjugate foci*, and are related by a very simple rule: if the distance of one is *u* and of the other *v* and the focal length *f*, then *v* is the product of *fu* divided by their difference, *u* being greater than *f*: if it is less, then *v* is the distance beyond *u* from which the rays appear to diverge after passing through the glass.

The focal length of a glass lens bears a very similar relation to the radii of curvature of its two surfaces. For it is twice their product divided by their sum; or by their difference, if one side is concave but the lens effectively convex or thickest in the middle, which is called a *meniscus*. Therefore the focal length of an equi-convex lens is the radius of curvature of each face,

and of a plano-convex it is twice the radius of curvature of the one spherical face. Another practical rule
is that the focal length is the square of half the width
of the glass divided by its effective thickness or excess
of the middle over the edges. But these rules only
hold for glass, whose refractive index is ³⁄₂.

Concave lenses also have a focal length, which may
be calculated by the same rules, but it means a different thing. When parallel rays fall on such a lens they
are made to diverge as if from a point much nearer
than the object, and the distance of that point is called
the focal length. Rays really coming from that point,
or any other, by no means emerge parallel, but are
spread out wider by a concave lens.

A pencil of rays may pass either ' centrically' through
a lens, like the pencils A O*f* and all the others through
the object glass of the telescopes (p. 312), or ' eccentrically ' as they do at *a* and *b* in the eye-glass. The
surfaces across the centre of a lens are practically parallel to each other, or perpendicular to the axis of any
pencil of rays through the centre; and therefore centrical pencils are not refracted at all, though the rays
which compose them are refracted to a focus somewhere in the axis of the pencil. But eccentrical pencils are wholly refracted, and bent inward, and so made
to converge to the eye E beyond the lens.

That also shows how a common magnifying glass or
convex lens magnifies things near the focal distance.
All the points in the object, except the middle one, are
seen by eccentrical pencils, which are refracted and
made to converge to the eye at wider angles with each

other than the pencils by which those points are seen without the glass. Without the glass the apparent size of ♂ would be ƒE*i* (drawing lines ƒE, *i*E to the eye); with the glass it is *a*E*b*, a much larger angle. If the object is much nearer the glass than its focal length, the rays of each pencil diverge too much for the glass to make them parallel enough for distinct vision; and besides that, the angle ƒE*i* is then nearly the same as *a*E*b*, and so there is little magnification. On the other hand, if it is too far off, the rays of each pencil converge into the eye and again produce indistinctness; and the pencils also become so nearly centrical that there is no sensible magnifying. The magnifying power of a glass for objects near its focus evidently varies inversely as the focal length.

Images.—But a magnifying glass has another and a very different effect, much less known, on which the action of a telescope depends. If you fix a common long-sighted spectacle glass in the open window, and recede from it, any object which you see through it becomes more and more confused, until it suddenly reappears inverted and smaller. In fact it is no longer the object itself that you see, but its image formed as I described at page 303. If the object is so far off that the rays of each pencil may be called parallel, the image will evidently be at the focus of the lens; and the width of the image is to that of the object as the focal length is to the distance, or = the numerical value of the apparent diameter × the focal length. If the object is a little beyond the focal length, the image is farther off and larger, as in microscopes. So the transparent pictures put upside down into a magic lantern, a little beyond the focus of a convex lens, cast a magnified image of themselves upright on the wall, or on a wet sheet if you want to see them from the other side. On the other hand the small bright spot under a burning glass is the image of the sun, taking his shape reversed in an eclipse, or that of the crescent moon.

Spectacles.—Short-sighted eyes require the rays from each point to come in a diverging state, as they do when the object is very near, though the pencils from the different points then come more converging, and so the object looks larger than average eyes can ever see it. Therefore concave spectacles suit short sight, which make the rays diverge. Long-sighted eyes require parallel rays, and therefore cannot distinctly see things near, without convex glasses to make the diverging rays parallel, not really to magnify the objects, though they do so. Holding a book far off makes the rays of each pencil come parallel or nearly so, but it also makes the letters appear smaller and send less light into the eye; so that it is the same as reading smaller print by a worse light; therefore it is better to use spectacles.

When we talk of parallel rays, it always means the rays of one pencil and not the rays of different pencils; only the stars are so far off that the pencils of rays from their outsides come to us at no angle that can be measured by the utmost magnifying power, and therefore practically as parallel as the rays of each pencil; and that is why a star in the largest telescope looks nothing but a point, though Neptune or the satellites of Jupiter can be magnified into discs of visible size. For the rays of a pencil diverge no more than the width of the *glass* divided by the distance, while the extreme rays or pencils from a planet make an angle = the width of the *planet* divided by its distance.

Astronomical Telescope.—You can now understand the operation of an astronomical telescope, and how its glasses must be placed, and how much it magnifies. The image being formed at the focus of the object glass differs from an ordinary small picture in the fact that the rays can only go from it in the directions in which they are sent by the object glass. Consequently the eye glass must be exactly at the proper place to receive them and bring them in a proper state for vision to the eye : that is, for ordinary eyes it must have its own focus coinciding with the image or the focus of the object glass, and be moved a little nearer to it for short sighted eyes which require diverging rays.

Magnifying Power.—Then for the magnification : the pencils aE, bE are parallel to fc, ic, the lines from f and i through the centre of the eye glass, because f and i are at the focal distance ; and so the apparent size of the image aE$b = fci$, instead of AOB the naturally apparent size of the object, which is evidently the same as fOi. Those angles being small are inversely as the distances of the image fi from O and c (the angles are immensely exaggerated in the figure for distinctness) ; or the magnification is the focal length of the object glass divided by that of the eye glass. And that is the case in all telescopes, except two of the reflecting ones which I shall mention afterward.

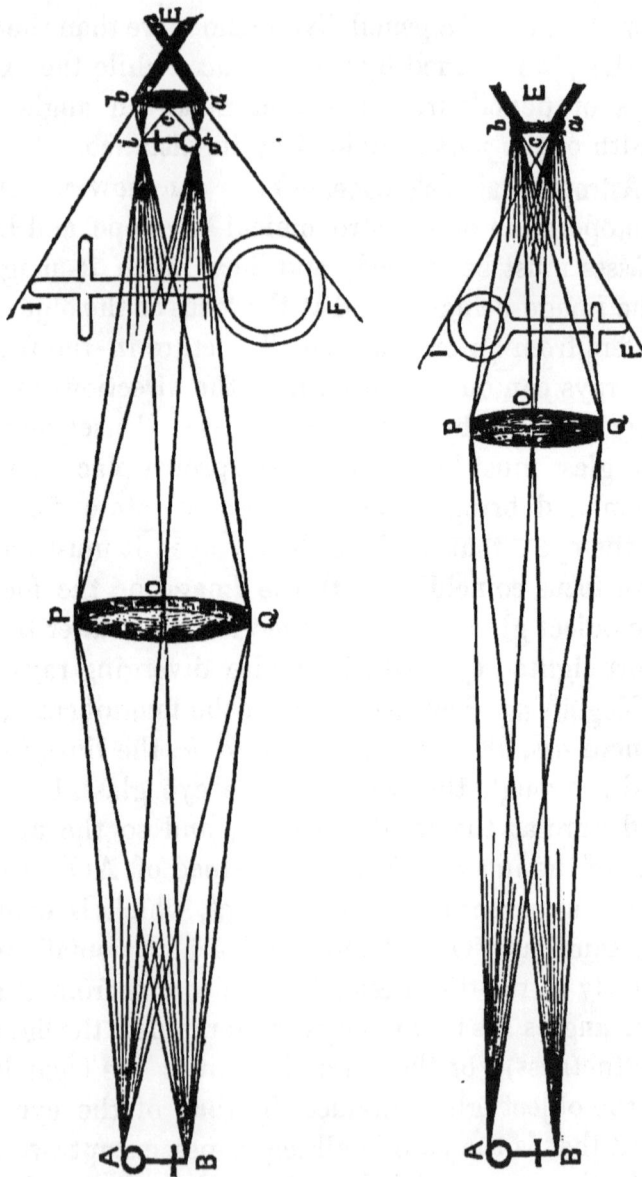

You see then that the business of the object glass is not to magnify the object, but to form an image of it to be magnified by the eye glass; and the flatter the object glass (provided it has some convexity), or the greater its focal length, the larger is the image that it forms. The average diameter of Jupiter being nearly 40″ or ·0002, the size of its image from a glass of 10 feet focal length is ·024 inches; and with an eye glass of 1 inch focal length, the apparent diameter of ♃ will be magnified 120 times, or into 80′ or nearly 3 times that of the sun or moon, and Jupiter will look 7 times as big as the moon, and Saturn nearly as big as the moon.

But magnifying power is reckoned by diameter and not by disc, which varies as the square of the diameter.*

Galileo's Telescope (see figure at p. 312) has a concave eye glass placed at its own focal length before the focus of the object glass instead of behind it. Therefore it makes the rays of each pencil come out parallel, but the pencils themselves diverge into the eye at an increased angle *bca*, instead of converging to E. But that produces a widened image in the eye just as well. Here no image is formed in the telescope, and there is no inversion; for though the pencils A*a*, B*b*, cross at O as before, the A pencil from the top of the object comes into the eye in the direction I*ca*, as from the top of a larger ♀, and the B pencil F*cb* comes from the bottom. The crossing of the apparent directions does

---

* This does not apply to advertisements of microscopes magnifying drops of water many thousand times: they evidently call a magnification of two hundred, 40,000.

14

not signify. In the astronomical telescope the rays cross in a different way at E where they enter the eye, and a *diaphragm* or plate with a hole in it is fixed there to guide the eye to the proper place.

Field of View.—But the capacity or field of view in the Galilean telescope is limited by the size of the pupil of the eye, as the pencils diverge into it; and for the same reason very high magnifying powers cannot be used. I have drawn all the parts of both telescopes out of all real proportion, or else the small angles made by the rays would not be visible. The field of view AOB, or the largest apparent diameter that can be seen, evidently = the angle aOb or the width of the pupil divided by the length of the Galilean telescope, even if the eye is close to the eye glass: which is a reason for keeping them short. In the astronomical telescope that angle is not the width of the eye, but the width of the eye glass, divided by the length of the telescope, measuring by the axis of the two extreme pencils, which however allows half of their rays to slip over the edges of the eye glass. And the field is limited to that by a diaphragm fixed across the focus to stop any pencils more oblique than those whose axes fall just within the eye glass. Glasses of higher power or shorter focal length are also smaller, and so the field of view is less the higher the power that is used in the same telescope.

Long telescopes with high magnification often have a *finder* set upon their back, which is a short and small telescope of less power, in which the eye glass is much larger in proportion to the length, and therefore the field of view greater. This also was an invention of Newton's.

Another serious defect of the Galilean telescope for astronomy is, that, as the rays never reach a focus, there can be no cross wires set there to mark the transit of a star, or to define its exact position, which is a most important part of the business of telescopes. The inversion by the astronomical telescope does not signify. In land telescopes of that construction another lens, or more, is introduced to reverse the image back again, and then it is far better than the Galilean.

Brightness. — Although a telescope gathers much more light than the eye, and would act as a burning glass, it does not make objects look brighter, but dimmer. For if all the light received by it enters the eye, it can only exceed that received without the telescope as the area of the object glass exceeds that of the pupil, or as the squares of their widths. And if you draw a pencil of rays going centrally through both glasses you will see that its width at the object glass (which is the width of the glass itself) is to its width at the eye glass (and therefore to that of the pupil, if the whole pencil is to enter it) as their focal lengths, or as the magnifying power. But the rays are spread out wider and thinner over the magnified object or image in proportion to the square of the magnifying power, and therefore (under the above condition) as the area of the object glass is to that of the pupil. So the magnifica- · tion diminishes the brightness at least as much as the size of the object glass increases it ; and the brightness is reduced still more, either if the pencil does not fill the eye by reason of the magnification, or if it is too large to be all taken in. Beside that, there is the loss of light in the glasses, as stated at page 298. So the image in a telescope can never be as bright as the object.

Color Dispersion.—But both kinds of refracting telescopes, so far as I have described them yet, have a serious defect, which made Newton and others abandon them for reflecting ones. If you look through a strong magnifying glass, or a common cheap telescope, you see things fringed with colors and indistinct; as is still more visible in the magnified images of a magic lantern. The reason is that the seven colors which make up a ray of white light are not refracted equally by any substance; or every ray of white light is split up by refraction into seven colored ones, each of which will not divide again. It is not very easy to see it at one refraction, but it is evident enough with two, as when a ray emerges from a prism. Turning back to p. 302, the whole emergent ray does not come in the direction bG, but only the middle or green part of it, while the red is refracted through the smaller angle ZbR, and the violet as much as ZbV: the whole spectrum consisting, first of invisible cool and chemically acting rays, most refracted, then violet, indigo, blue, green, yellow, orange, red, and then invisible hot rays, as described at p. 81. Those at the violet end are comprehensively called blue, and those at the other end of the spectrum, red.

Consequently a lens cannot bring any pencil of rays really to a focus; but a blue image of every point of the object is formed nearer to the lens than the red one, and a green one somewhere between the two, which is overlapped by the red rays before they reach their focus and by the blue diverging again after they have passed their focus. Therefore the smallest image of each point in the object is a little circle about half way between the red and the blue foci; and that circle is never less than one-fiftieth of the width of the lens. Consequently large object glasses could not be used until this defect was cured by the invention of compound *achromatic* or colorless glasses, which I will describe presently.

Not only does the space-penetrating power depend on the size of the object glass, but the power of ' dividing ' double stars, in consequence of the 'spurious disc' being less with a large glass than a small one (p. 282 ; and see R. A. S. Monthly Notices, 1867, p. 235).

Moreover the color dispersion is increased to the eye by increasing the magnifying power of the eye glass, or using one of very short focal length ; and as the magnifying power of the telescope is the focal length of the object glass divided by that of the eye glass, the only way of getting a highly magnifying telescope was to have an object glass of very great focal length. Accordingly they were sometimes made from 100 to 600 feet long, so that the object glass had to be set upon a pole, without any tube, and with strings and levers to pull it into the right direction for sending the rays into the eye glass. Such telescopes were called aërial. We are returning to them in one respect; for tubes are

being superseded by open frames for very large telescopes, but the glasses are of course fixed in them.

Achromatic Telescopes.—By an unlucky mistake in
an experiment Newton missed the discovery which was
not made use of till nearly a century afterward, by
Dollond the optician, though it was made before by
Mr. Hall of Worcester, that glasses of different densities have different powers of dispersion into colors for
the same amount of general refraction. A lens of *flint
glass*, which contains lead and has a specific gravity
3·6, spreads the blue rays farther away from the red,
though it may refract the red or the green equally with
a crown glass of the same convexity, whose specific
gravity is only 2·5. Therefore a slightly concave flint
lens may refract the blue rays outward just enough to
correct the excessive refraction of them inward by a
very convex crown lens, while it leaves a general balance of refraction over, because the two lenses together
are on the whole convex or thickest in the middle. Or
the same thing may be done by putting a double concave flint lens between two convex crown lenses. I
have drawn the object glass of the Galilean telescope
in this way, and that of the astronomical with only two
glasses, in the pictures at p. 312.

The rule for determining their proportions is simply
that the focal length of the flint glass must bear the
same proportion to that of the crown glass (or of the
pair of them) as their respective dispersive powers, or
as ·068 to ·033. For the width of the spectrum of a
flint glass prism, or the angle between the extreme red
and violet rays, is ·30 of its mean angle of refraction,

or the refraction of the green ray, while the ratio of dispersion to mean refraction by a crown glass is generally ·17, or about half as much as by the flint glass.

Secondary Colors.—Still the correction is imperfect, by reason of what is called the *irrationality* of dispersion ; which means that the ratio of dispersion to the mean refraction is different for the different colors by any two kinds of glass. A flint glass prism which disperses the blue rays as far from the red as a given crown glass (of a greater refracting angle) does not disperse the green rays quite as far from the red as the crown glass does. Consequently when the prisms are put together, turned opposite ways, they bring out the red and blue rays together, but leave a little green behind ; and so the image through a compound lens of that kind is still fringed with green, and has not a sharp outline ; or the *definition* is imperfect. And although this secondary dispersion is only a 60th of the primary, and therefore a 3000th of the width of the object glass, it is enough to leave a sensible advantage on the side of reflecting telescopes, which make no colors. Even if a third glass of some different composition is used, none has yet been discovered which completely annihilates the colors.

Dr. Blair indeed did effect it in a temporary way by fluid lenses (of course enclosed between glasses), which had the requisite action on the different colors : but they could not be made to last. Certain oils and gums refract green light much less than even flint glass ; and a long table of them, in the inverse order of their effect on green light, is given in Sir D. Brewster's treatise on

Optics in the Encyclopædia Britannica. Mr. Grove 17 years ago made some solid lenses of resin and castor oil,* and also of hardened Canada balsam and castor oil, interposed in a meniscus form between the crown and flint glass lenses, which he said very nearly destroyed all secondary color; and he advised opticians to pursue the subject on that footing, as they alone have generally the means of doing it. Mr. Wray, an optician, has lately *patented* a 'semi-fluid cement' of oil of cassia and castor oil, or Canada balsam, to be interposed as a meniscus lens, 'hermetically sealed' between the crown and flint lens, the crown glass being made flatter at the back than usual, to leave room for the meniscus of cement, as Mr. Grove's was.

This is said to answer extremely well for the achromatism, but great doubt is expressed as to the possibility of making any but a solid composition permanent, under the changes of temperature to which telescopes are exposed. For an observatory must keep nearly the same temperature as the air outside, or the refraction is disturbed (p. 215). The glasses of a compound object glass are indeed often united by a transparent cement of the same refractive index as glass, as nearly as possible; since that prevents internal reflection (p. 305), by making the two glasses as it were continuous, instead of leaving a film of air between them. But such cement is of the same thickness throughout, and as thin as possible; while these dispersive cement lenses are thickest in the middle, and therefore do not expand and contract equally at the middle and the

---

* See Ast. Soc. Monthly Notices of 1853, and Phil. Mag., March, 1867.

edges, and are difficult to keep hermetically sealed between glasses if they are fluid. The principle of these inventions is, that the oils and resins refract green light less than glass ; and therefore if the convex lens is in effect made partly of glass and partly of these things, they produce together a convex dispersion of all the colors in such a ratio that the concave dispersion of the flint lens can correct it : or that the two opposite dispersions become ' rational ' instead of irrational.

Spherical Aberration.—Independently of color dispersion, a lens with spherical faces does not bring the rays of any pencil or of any color absolutely to a point ; for those which come through the edges are brought to a focus rather sooner than those near the middle of the lens. The amount of this spherical aberration, as it is called, depends on the shape of the lens, and even on which side of it faces the light, if the two sides are different; and it is always much less than the chromatic dispersion even of a single lens. The shape which gives the least aberration (of those which can be used) has the front face 6 times as convex as the back; which opticians call a *crossed lens.* The diameter of the least circle of aberration, or the image of a point in such a lens, is only the 2160th of its width, when the focal length is 10 times the width. And it increases as the cube of the width, but inversely as the square of the focal length ; so in that respect also the long aërial telescopes had an advantage.

Theoretically this defect may be cured by making the faces of the glass not quite spherical, but practically
14*

they cannot be ground in any other form. And fortunately the aberration can be corrected in another way,
by selecting proper degrees of curvature for the separate
lenses of a compound object glass, without interfering
with their achromatism, as that depends only on their
focal lengths, not on their total convexity and con-
cavity. Various distributions of the curvature will
answer, and the Rev. C. Pritchard, P.R.A.S., has calculated a table of corresponding curvatures to suit various focal lengths, so that any glasses may be matched
at once. The best general forms are like those in
which I have sketched the two object glasses at p. 312
with the curvatures very much exaggerated. The
crown glass in front has its front face more convex
than the back, and the flint glass behind fits it, and has
its back convex, but less so than either of the crown
glass faces, so as to be effectively concave (which is
called a 'reversed meniscus'), while the whole object
glass is of course effectively convex. It is not worth
while to go through the similar arrangements for treble
glasses.

Compound Eye-pieces.—Hitherto I have spoken of
the eye-glass as a single lens : which it still is for very
high magnifying powers. But where distinctness or
definition is chiefly aimed at, the eye-glass also must
be corrected for color, for spherical aberration, and for
something else. For suppose a centrical pencil of rays
to be brought exactly to a focus by the object glass :
it goes on to the eye-glass eccentrically, except of
course the one pencil which keeps the axis of the telescope ; and then instead of being refracted into a

round and colorless stick or pencil of parallel rays, it is spread out into a colored oval one, of rays not parallel in one direction (that of the plane of the paper in the figures at p. 312) but diverging. Moreover all the points of the image ought to be at the focal length of the eye-glass, or it ought to be concave toward the eye-glass: but in fact it is rather concave the other way, being formed by the object glass. Consequently the rays from the outside of the image come from rather too far off and are more turned inward by the eye-glass than if they came exactly from its focal distance.

It is found that the best way to correct all these errors is to divide the work of the eye-glass between two less convex lenses separated by a space equal to the mean of their focal lengths, or 2 inches if the focal length of the first is 3, and of the second (next the eye) 1 inch. The first is called the field glass, because the field of view now depends on that (see p. 314), and it stands at half its own focal length before the focus of the object glass; so that the image is now formed between the field glass and the eye glass. The best shape for them is a *meniscus* field glass with radii of curvature as 4 to 11, and a ' crossed lens' (*i.e.*, of radii 1 and 6) for the eye-glass, both with their convex faces toward the light.

This is Huyghens's eye-piece, except that he had two plano-convex lenses; and it is also called the *negative* one, because the image is behind one glass, but before the other. And this involves a serious defect for some purposes. The image no longer has the same proportions as the object, but the outer parts of it are con-

tracted, because the pencils near the edges of the field
glass are most refracted or bent inward. The image
of a square would thus have its corners bent in or the
sides curved out, and a piece of netting would appear
to have the outer meshes smaller than the middle
ones. Consequently it will not do to put cross wires
there for measuring an object or its position.

For that purpose the *positive* eye piece was invented
by Ramsden, a famous instrument maker about 1780.
In it both the glasses are beyond the focus of the object glass, where the wires are, and so they and the
image are equally magnified by the eye piece. The
glasses are two equal plano-convex lenses with their
convexities facing, at a distance $= \frac{2}{3}$ of each of the
focal lengths; and the field glass is a quarter of its own
focal length beyond the focus of the object glass. It is
not quite achromatic, but it has less spherical aberration and distortion than the other.

The principle of both the positive and negative eye
pieces is this. An achromatized pencil from the object glass reaching the field glass eccentrically, its blue
rays are there most refracted, and are sent to the eye
glass nearer its centre than the red rays which are
least refracted. But those which strike the eye glass
nearest the centre, or least obliquely, are least refracted
by it; and so the less refraction there balances the
greater refraction before, and the blue and the red
come out parallel, and then the lenses of the eye unite
them into a colorless point. The same kind of compen-
sation takes place also between the rays which are most
refracted and least refracted by the field glass by reason

of spherical aberration, and the convexity of the image being turned the wrong way; and so these compound eye pieces correct both defects.

The inverted image is turned into an erect one in land telescopes by an eye-piece generally made of four glasses; but we are not concerned with them. All eye pieces are made adjustable for different eyes by the eye glass sliding nearer to the object glass for short-sighted eyes, which require the rays of each pencil to diverge a little instead of being parallel.

Micrometers.—The measuring wires which I have several times spoken of are stretched across a small frame which slides a little across the telescope with a screw, of which the head is graduated, so that you can measure exactly how far any wire has to move to bring it into contact with a star, to measure its distance from any other star or the meridian or zenith. The same screw apparatus is also used outside some telescopes to adjust or measure their position, with microscopes attached to it, and in each case it is called a micrometer, or measurer of small distances. The wires are fixed in and parallel to the meridian for transit purposes, and horizontally for measuring altitudes and polar distances. The wires in the reading off micrometers are set so that they may be brought to coincide with the graduations of the circle they are to read off.

Since each pencil of rays spreads over the whole object glass, half a lens makes the same image as the whole one, only not so bright. If the two half lenses slide along the common diameter two images will be visible whenever the two centres do not coincide.

And the distance from the first contact of the images to the last, measured by the turns of the screw which moves one half lens along the other, is evidently twice the width of each image. This makes the best micrometer for measuring diameters of the sun and planets and distances of double stars, and is called a *heliometer*.

In order to escape from the inconveniences of either color dispersion or long aërial telescopes, Newton and Gregory and Cassegrain and Herschel invented their several forms of the reflecting telescope, which produces no colors, except in the eye glass, which Huyghens had corrected long before compound object glasses were made. And they are still used because mirrors or *speculums* can be made much larger than object glasses in the present state of art; and I have shown you already that the space-penetrating power of a telescope depends on its size, or the quantity of rays it can take in.

Herschel's largest telescope had a speculum 4 feet wide, and Lord Rosse's 6, with 53 feet focal length weighing 4 tons; whereas the largest lens that has yet been *completed* is only a little more than a foot wide, with a focal length of 12 feet. So that it greatly exceeds any refracting telescope both in penetration and magnifying power; for in reflectors as well as refractors the magnification varies as the focal length of the object mirror. But a telescope is now in progress of which the compound object glass is 25 inches wide.

It has been cast by Messrs. Chance of Birmingham and made into a telescope by Mr. Cooke of York, and thus far there is every reason to believe it is successful. It has a steel tube 30 feet long.

The theory of reflection is much simpler than of refraction, and is all deduced from this one law of nature, that rays are always reflected from any surface at the same obliquity as they fall upon it. Consequently an object appears as far behind a flat mirror as it really is in front of it, the apparent distance being determined, as the real distance is, by the different view which each eye takes, or else by moving one eye (p. 102). Therefore you can see your whole body in a mirror half as wide and half as high as you are.

Reflection from curved surfaces is not so simple as from flat ones; but, as with lenses, it is much simplified by the mirrors of telescopes being only small segments of their spheres, and by the rays only falling on them at a very small obliquity. I had occasion to tell you at p. 242 that the tangent at any point of an ellipse (or a straight line which coincides with the ellipse there for a moment) makes equal angles with the two focal distances SP, HP (p. 238). Consequently by the law of reflection rays of light emanating from one focus of a prolate spheroidal mirror are all reflected to the other. Again a parabola is only part of an ellipse with one focus at an infinite distance from the other (p. 258); and therefore rays coming from an infinite distance, *i.e.*, parallel rays, are all reflected by a parabolic mirror to its focus; and rays from a lamp at the focus are all reflected parallel to the axis. Moreover it may be proved that a parabola for a short distance near the apse coincides with a circle whose radius is twice the distance of the focus from the apse; and therefore a spherical mirror reflects parallel rays (not far from its axis) to a point half way along the radius of curvature, subject to the spherical aberration, which is due to the slight deviation of the circle from the parabola.

Fortunately that is much less than from a lens of the same size, as there are no means of correcting it. The diameter of the least circle of aberration or smallest image of a point is the cube of half the width of the mirror divided by the square of 4 times the focal length, or by the square of the diameter of the sphere of which the mirror is a part. So that it is only the 12,800th of the width of a mirror whose focal length is 10 times its width; and in Lord Rosse's 6 feet speculum of 53 feet focal length the image of a point is a spot ·006 inches wide, subject to magnification by the eye glass of the telescope.

A concave mirror then deals with rays which fall upon it like an equi-convex lens whose radii of curvature are half that of the mirror, or a plano-convex of the same curvature as the mirror; but that is only true of glass whose refraction is $\frac{3}{2}$ (p. 308).

All reflecting telescopes have a concave mirror as wide as the tube, placed at its lower end, which receives the rays from the object and sends them to a focus, and forms an image, exactly like an object glass, only without colors. Its diameter again = the focal length × the numerical value of the angle representing the apparent diameter of the object, or its real diameter divided by its distance. Thus the image of Jupiter in Lord Rosse's telescope is about an 8th of an inch wide ; and the spherical aberration or image of each point is spread over a 20th of the width and a 400th of the area of the whole image, whether it is magnified much or little by the eye glass. You see this is considerably less than the chromatic aberration of a lens only a foot wide with 10 feet focal length, even when it is corrected as much as possible (p. 319). A concave mirror, like a convex lens, magnifies things if your eye is near it, but diminishes and inverts them if you are far enough off to see the image and not the object in the glass : a convex mirror, like a concave lens, diminishes without inverting.

There are four ways of dealing with the rays after reflection from the great mirror. In Gregory's telescope they were allowed to come to a focus, making a first image there, and then were reflected again by a smaller concave mirror set in the middle of the tube to another focus which is also the focus of the eye glass, passing to it through a hole in the middle of the great mirror; and so the first image is reversed back again at the second. In Cassegrain's telescope the second mirror is convex, and receives the rays before they come to a focus, acting like a concave lens, only sending the rays backward instead of forward, less converging than before, to a focus which again coincides with that of the eye glass.

In Newton's telescope the second mirror is a flat one, which makes no difference in the convergence, and also receives the rays before they reach a focus, and might send them to the eye glass through a hole in the great mirror like those other two telescopes. But instead of that, the small mirror is set obliquely in the tube, at an angle of 45° to the axis, and so it reflects the rays out sideways at right angles to the axis. Herschel's telescope dispenses with the second mirror altogether; for the great one is set a little askew at the bottom of the tube and sends the rays to an eye piece fixed just within the edge of the tube at the other end. This can only be done in very large tubes, where the observer's head in the mouth of the telescope only cuts off a comparatively small part of the rays.

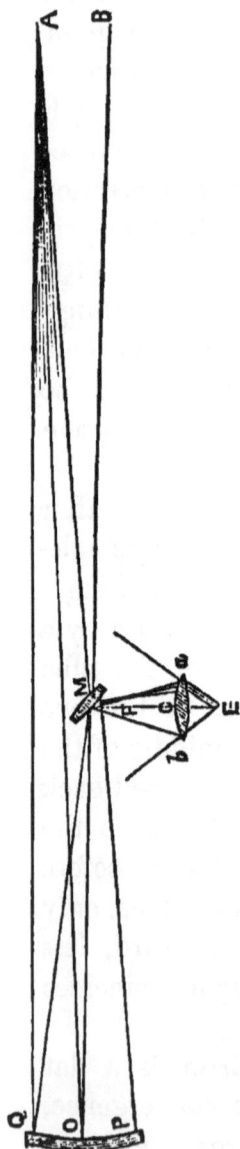

The action of Newton's telescope is shown in this figure, in which you may easily trace the rays and pencils of rays from the points AB of the object to the large mirror OPQ and the small one M, and thence to *a* and *b*, converging to E the eye at the wide angle *aEb*. Cassegrain's telescope has one advantage over Newton's, that the small mirror being convex tends to correct the spherical aberration of the large one. Gregory's on the contrary, being concave, aggravates it. The magnifying power of them both may be said to be the square of the focal length of the great mirror divided by the product of the focal length of the small one and of the eye glass. The magnifying power of Newton's and Herschel's is the same as in the two refracting telescopes, *i.e.*, the focal length of the great mirror divided by that of the eye piece.

Sir J. Herschel expects the Newtonian telescopes to supersede all others now that a method has been invented, which Newton himself again divined as possible, of silvering concave glass speculums instead of casting those heavy metal ones which also have to be reshaped every time they are repolished. Lord Rosse's speculum weighs 4 tons and is supported at the back on many points by a complicated system of levers to balance the pressure; for an inequality of the thickness of a thread bends a heavy speculum enough to destroy all convergence to a focus. It is not the quicksilver back of the glass that reflects, as in a looking glass, but a thin film of silver chemically deposited on the front or hollow face, which can be repolished, and even renewed when necessary, the glass being once for all ground to the proper shape. If this can only be made parabolical, so as to get rid of spherical aberration, the days of refracting telescopes probably are numbered. It is said that a good speculum of this kind reflects nearly as much light as is refracted through an object glass, while a bell-metal speculum loses one-third of the light. The only difference be-

tween speculums and bells is that bells are (or should
be) made of 13lbs. of copper to 4 of tin, while speculums
are 128 to 59, for reasons given in the Treatise on
Clocks and Bells.

All the reflecting telescopes but Gregory's make an
inverted image at the focus of the eye piece; and they
all require compound eye pieces like refracting telescopes. Newton's has this farther advantage over all
other telescopes, that the reflection out sideways may
be made to take place through the hollow cross axis on
which the telescope is balanced (if the great speculum
is not a heavy one) and so the observer may sit still in
one position and has neither to look upward, which is
fatiguing, nor to mount into a box high in the air to
look downward as with Herschel's. Lord Rosse's
telescope is on the Herschelian principle.

Gregory's and Cassegrain's telescopes are almost
abandoned because of the quantity of light which they
lose by the second direct reflection. Much less is lost
by the oblique reflection in Newton's, especially when
a reflecting prism is substituted for the second mirror,
as explained at p. 305.

Helioscope.—There is yet one other kind of telescope
which ought to be described now that observations of
the sun's surface are becoming an important part of
astronomy. The sun is far too bright to be examined
through any common telescope; for though it does not
increase the apparent brightness, as I explained at p.
316, still it acts as a burning glass concentrating all
the light from the object glass into the eye. Consequently it is necessary to reduce the quantity of light

very much, and yet to keep the other advantages of magnification and of gathering a large pencil of rays from every point. For it does not answer to reduce the light by using a very small object glass, or covering up all of it except a small hole in the middle. It is necessary first to gather a large bundle of rays, and then as it were to filter them by some contrivance which will only let a small part of the light come into the eye. The common expedient of smoked or colored glass will not do either; for such a glass soon gets heated by the rays which it stops, and a colored image is an imperfect one.

The plan generally adopted is to make them fall obliquely on one of the wide faces of a thin prism, a little before they reach the focus. That is a very different thing from sending them directly into a narrow side of a right angled prism, as in the Newtonian telescope, which is intended to reflect them all internally from the wide face. This thin prism reflects a little of the rays from its first face, by virtue of the property which all transparent substances have of reflecting some of the light which falls on them while they transmit all the rest. Here about one 30th of the rays are reflected, and the rest refracted into and through the prism, and sent away as not wanted through the end of the telescope which is left open, while the reflected rays come out sideways into an eye piece as in Newton's. The reason for using a prism instead of a flat piece of glass is that the internal reflection from the second surface also would interfere with the reflection from the first surface if they were parallel, whereas the prism sends it off in another direction.

This gathering and filtration of the rays is sometimes carried farther, by using a large double concave object glass, unsilvered, as a reflector of some of the rays which fall on its first surface, leaving all the rest to be dispersed through it into the air behind. The rest of the construction is the same as before. But M. Foucault, the inventor of the pendulum experiment for proving the earth's rotation (p. 33), has lately discovered that an object glass covered outside with a thin film of silver or gold burnished bright will reflect away the greatest part of the rays, while it is transparent enough to transmit some, which are fit for vision and examination of the sun, but only slightly colored blue. Probably this helioscope will supersede the others, except that it requires a distinct object glass and practically a distinct telescope, while the first only requires a prismatic eye piece to be used instead of the common one, which is movable.

We have yet to consider the different ways of fixing telescopes for different kinds of observations.

Transit Circle.—This most important telescope is mounted on a cross axis lying east and west, so that the tube can only move in the plane of the meridian. It is now made so as to combine the work for which two telescopes were formerly used, one called a *transit instrument* for observing transits across the meridian, and the other called a *mural circle* because it was set against a wall for observing distances from the equator or the pole. The Greenwich transit circle tube and its cross axis are of cast iron, instead of brass as usual. The axis or its pivots until lately were always set on what were called Vs, being bearings of that shape ; for a pivot cannot be made to lie steady in a semi-circular bearing. But at last this rude method of obtaining a steady bearing has been superseded by the simple device of cutting a wide notch or piece out of the bottom of a semi-circular block, so that the pivot still presses only on the sides, but with a much wider bearing than on Vs, which the axis only touches in a line on each side, and often wears away unevenly.

Moreover the new compound of copper and aluminium, which will probably be called *al-bronze* (as bronze is a softer bell metal), is coming into use for telescope bearings, as it is four times as strong as brass. It was hoped that it would do instead of silver bands or circles for graduation ; but though it spoils with the air less than brass, it is far inferior to silver. It is evidently essential to accurate observation that the bearings should be not only level, but as firm as possible, and they are consequently laid on piers built deep into the ground. The instrument is used to observe the transit of a star across the meridian, by pointing it to the proper height to catch the star, and then looking at it as soon as it comes into the *field of view*, and observing the time by the clock when the image of the star crosses the middle micrometer wire, or rather all the wires in succession. The difference between the times of two stars crossing the meridian is the difference of their right ascension, which is now generally given in tables of stars in time and not in degrees (p. 292.)

Again, distances from the zenith are found by measuring the angle which the telescope makes with the
vertical when a star on the meridian is seen exactly on
the wire which is set across the focus horizontally, or
at right angles to the transit wire. And that is done
in a remarkably neat way, by first looking at the star
itself, and then at its reflection in a basin of mercury
which forms an artificial horizon : sometimes a plate
of glass is laid upon it to stop any tremor of the surface
and to keep it clean. By the law of reflection (p. 327)
the telescope always makes the same angle with the
horizon when pointed at the star and at its reflection,
only in one case it is looking upward and in the other
downward. Therefore the altitude of the star, or 90°
—the zenith distance—is half the angle moved through
by the telescope between the two observations. And
the declination, or distance of the star from the equator, or what is more commonly used, its north polar
distance (N.P.D.) is easily got from its zenith distance
and the colatitude of the observatory, which is the constant distance of its zenith from the pole, as the latitude
is the distance of the zenith from the equator.

For this purpose the transit circle has a large wheel
fixed to one end of its cross axis with a graduated rim,
and the graduations are read off by several of those
microscopes called *micrometers* (p. 325) which are fixed
at convenient places on the pier near the rim of the
circle. In the old *mural circles* the graduated circle
is fixed on the wall, and the telescope carries an index,
with micrometers for reading off accurately. The
telescope need not be so accurately in the meridian for

observing N.P.D. as for transits ; and so setting it against a wall as near N. and S. as possible was enough without all the precautions which a transit instrument requires. For [the bearings of a transit instrument must be exactly level and of the same size, and their centres exactly east and west, and the pivots which ride in them exactly round and exactly alike, and the cross axis exactly at right angles to the telescope axis, or the line through the centres of all the glasses. These things require constant watching and testing, and mathematical formulæ have to be calculated for the necessary corrections. Most of them can be tested by lifting the axis over and reversing the pivots, and looking through the telescope opposite ways ; but for heavy cast iron tubes other methods equally accurate are used at Greenwich. These details however are beyond the scope of this book.

Altazimuth.—A transit circle being only able to move in the meridian of course can only see stars when they come close to the meridian. But if the horizontal cross axis of the telescope is carried by a frame which itself turns round on a vertical axis, or on a horizontal bed, it can be directed anywhere ; and it will measure altitudes above the horizon by moving the telescope up and down, and azimuths by turning the frame ; and so that is called an *Altazimuth*. The level plate at the bottom of the frame is graduated, and there is a fixed index with micrometers against it to show the degrees of azimuth through which you turn it.

The *zenith sector* is only another form, or rather a part of the same instrument, being a telescope moving
15

only over a moderately small graduated arc fixed at the zenith for observing stars near it, and therefore taking less room for its size than a complete altitude instrument with its vertical circle of degrees.

Equatorial.—Again if the frame in which the telescope is set turns on an axis not vertical, but parallel to the earth's axis, that also gives an universal motion; the telescope can be pointed to a star and then clamped or screwed fast to the frame, and if the frame is turned round from east to west at the same rate that the earth turns the opposite way, the star will stay in the telescope. It is called an *Equatorial*, because each point in it then moves parallel to the equator. That is the instrument used for gazing at the planets or moon, and · sweeping the sky for new planets or comets, besides other purposes independent of the meridian. Therefore the largest object glasses that can be made perfect are generally used for equatorial. Herschel's great telescope was set in a frame which also turned azimuthally; and Lord Rosse's larger one, though it stands between stone piers, is capable of some sideway motion so as to be used equatorially, and can be kept pointed to a star for some time by machinery for the purpose.

In those heavy telescopes the speculum end of the tube stays on the ground. The lighter telescopes which are mounted equatorially are balanced on their middle: and they are often connected with a peculiar kind of clock which turns the frame at the proper rate, and so keeps the telescope pointed to the star without the observer having anything to do except himself to follow it. These clocks do not go by beats of a vibrating pendulum, which would be magnified by the telescope into leaps of the star; but they have either a revolving pendulum, or balls like the 'governor' of a steam-engine, or some equivalent contrivance to make the motion continuous.

This plan has been adopted lately for recording observations at once.* Another continuous motion clock drives a barrel covered with paper, turning round (say) once in ten minutes, and at the same time advancing endways, by its pivot being made into a screw, so that a pencil fixed against it would draw a spiral line round the barrel. When the observer sees his transit he touches a pin, which by the common electric telegraph machinery makes a pencil strike the barrel and make a dot; which marks its own time, as the time at which every point on the barrel comes under the pencil is defined. The time of every observation, except mere gazing at the disc of a planet or moon or comet, is an essential part of it, and a clock is a necessary companion of almost every telescope; but clock making is farther removed from astronomy than telescopes, and I have written about that elsewhere.

There are many other minute contrivances for securing accuracy of position, observation, and measurement, which cannot be described here. You will have some idea of the accuracy required from this:—a good observer can tell when a star crosses the wire of his telescope within the tenth of a second by the clock, which is always placed so that he can both see and

---

* Appendix, Note XXVIII.

hear it, and count the beats from the minute and second which he has looked at.

And now that you have finished this book you must not fancy that you have learned astronomy: you have only been introduced to it, by a short and easy road. Even when you have learned all that is yet known of it, you may well remember the saying of Sir Isaac Newton, who at the end of his long life, after having opened more of the knowledge of the universe to human view than any man before or since, said that he felt like a boy who had been picking up a few pebbles on the sea-shore while the ocean of truth lay before him. He knew, like all true philosophers, lovers of true knowledge, that however far they range into the mysteries of nature, as into the 'star-sown deep of space,' they will find a wider space beyond, which they must wait to penetrate, until the time when 'no good thing shall be withheld' from those on whom the revelation already given has not been thrown away, and they shall no more 'see through a glass darkly.'

APPENDIX.

NOTE ON THE GREAT PYRAMID.

Besides the astronomical theory of the Great Pyramid, noticed at p. 52, the Astronomer Royal of Scotland propounds another, which is shortly this. The capacity of the famous porphyry coffer, which holds just 4 English quarters of corn and is certainly no mere stone coffin, expresses the earth's specific gravity 5·7 × a tenth part of the cube of a double cubit of 25·025 inches, the 20 millionth of the earth's axis (p. 15); and each side of the 13 acre base of the Pyramid was 365¼ such cubits.

That earliest and largest of all the Pyramids exhibits such proofs of mathematical design, and such astonishing precision of execution throughout, that we need not be surprised at any new geometrical relations of its parts being brought to light. But it requires an unusual amount of faith to accept this ingenious theory of Mr. Piazzi Smyth's on the present evidence. Nothing more is known of the 'sacred cubit' of the Jews than that it was somewhere about 23 inches, as Newton calculated. And if some of Noah's descendants were unconsciously inspired to choose exactly the 20 millionth of the earth's axis for their sacred cubit, no such measure appears in the Pyramid, except that two diagonals of the four corner-stone sockets cut in the rock are 100 and 200 inches. It is not to be found in the coffer, except by virtue of that hypothetical multiplier 5·7; which is not the earth's density according to any of the earth-weighing experiments (p. 32), or any probable mean of them. And if it is, how were the builders of the Pyramid to know it? Mr. Smyth admits only by revelation: a revelation of the earth's density 36 centuries before even the motion of the earth was known; unless that also was revealed and forgotten.

Lastly, the base of the Pyramid is itself still uncertain by at least two cubits, for reasons given in his book; and one unknown quantity can hardly be asserted with confidence to be $365\frac{1}{4}$ times another.

But some things in the Pyramid are known. Newton perceived, from the frequent multiples of 20·75 inches that occur in the passages and chambers, that that was the 'working cubit' of the builders. He did not know the capacity of the coffer, or he would have found that it just — a double working cubit cubed. And if he had known the angle of the slopes, $51° 51'$, he would have seen that it makes the width — $\frac{\pi}{2}$ or $\frac{11}{7}$ of the height, and the area of each face almost exactly — the square of the height. I wonder that John Taylor, who first perceived these things, did not see also that the builders having fixed on that slope, and being evidently fond of multiples of 10, were almost sure to take 440 working cubits for the base: that being the only multiple of 10 and 11 at all near that size, and making the height also a multiple of 10 and 7, or 280: all which figures stared them in the face by the rule on which they worked. And 440 such cubits, or 761 feet, are within a few inches of the same mean of the four latest measures of the base as Mr. Smyth himself adopts. I venture to propound this as a more probable theory of 'Pyramid metrology' than one which assumes a revelation of the density and diameter of the earth to Cheops or his architect.

NOTE I., PAGE 15.—Although national prejudice is plainly indicated by the author's language, his objections to the French metre are well grounded. But the advantages of a uniform standard of measurement are so great, that the French plan bids fair to become general, in spite of the inaccuracy of its basis.

NOTE II., PAGE 18.—Some recent observations on the temperature of deep wells have led the observers to suppose that the increase of temperature below the surface soon reaches a limit. Theoretical discussions of the tidal action of the sun and moon have also been published to controvert the hypothesis of a fluid nucleus.

NOTE III., PAGE 36.—The present estimates of the earth's mean distance from the sun vary from 91,000,000 to 93,500,000 miles.

Prof. Newcomb, of the Washington Observatory, has re-examined the observations of the transit of Venus in 1769, and after allowing due weight to more recent observations, he concludes that they indicate a distance of 92,383,000 miles, with a probable error of 136,000 miles. M. Liais, from the observations upon Mars in 1860, estimated the distance at 93,416,000 miles. Some considerations connected with the mean atmospheric inertia seem to indicate a distance of about 92,300,000 miles.

NOTE IV., PAGE 65.—The probable parallax fixed upon by Professor Newcomb is 8″·85.

NOTE V., PAGE 67.—Professor Wolff considers that Venus, the Earth, and Jupiter, all contribute sensibly by their attraction toward the formation of sun spots. I do not know on what ground the author asserts that the theory of a connection with magnetic disturbances seems to be abandoned.

NOTE VI., PAGE 67.—There is no sensible refraction in the atmosphere above the height of about 50 miles, but some meteoric and meteorologic observations indicate an aerial altitude of more than 200 miles.

NOTE VII., PAGE 80.—The experiments that indicate a greater velocity for electricity than for light, were conducted with wires, under circumstances that may possibly have been modified by induction. The *exact* coincidence of observed magnetic disturbances with disturbances in the solar photosphere, seems to render it probable that the normal velocity of light and electricity may be precisely the same. If a metallic wire could be made transparent there is no reason for supposing that it would transmit a wave of electricity more rapidly than a wave of light.

NOTE VIII., PAGE 83.—The nebular hypothesis assumes that the solar system was in a fused and vaporous condition before the "bringing together of its atoms commenced."

NOTE IX., PAGE 84.—If "heat and force are universally convertible," conversion is not the same as destruction. Matter and force are undergoing continual changes of form, but we have no reason for believing that either is ever destroyed.

NOTE X., PAGE 90.—Laplace estimated that the force of gravity must be transmitted with a velocity *at least* 6,000,000 times as great as that of light.

Note XI., Page 93.—The supposed eruption of Linné was not confirmed by subsequent careful observations, but other phenomena have been noticed upon the lunar surface which may be owing to volcanic action.

Note XII., Page 8.—The bright border that surrounds the "lumiere cendree," or darker portion of the moon's disc, in the new moon, has been attributed to refraction by the lunar atmosphere. Some of the photographs of the solar eclipses of August, 1868 and 1869, exhibit a degree of illumination of the moon's surface for which refraction would afford a ready explanation.

Note XIII., Page 93.—The town and building are supposed to be at the distance of 233 miles, and viewed in the same manner as the moon and Venus.

Note XIV., Page 96.—Sir John Herschel fancied that the lunar plains presented alluvial appearances, and Prof. Frankland thought that he could perceive traces of glacial action and moraines.

Note XV., Page 106.—The action of the full moon in dispelling clouds is still the subject of much discussion. The experiments of Melloni have established the fact that the moon's rays have a sensible amount of heat, but any lunar influence on the weather is probably more dependent on tidal attraction than on heat radiation. My own discussions of the rain-fall for forty years, at Philadelphia, and at Surrey, England, showed a tendency at each station to diminished rain-fall both at new and at full moon.

Note XVI., Page 116.—Confusion is sometimes occasioned by a failure to discriminate between the different relations of motion. The *actual* orbital motion of the earth at noon is in the same direction as the sun's *apparent* motion, and in an opposite direction to the motion of rotation at that time, or toward the point of compass that we then call west. But at midnight the motions of daily rotation and of annual revolution coincide in direction, each being toward the point of compass that we then call east. A stone let fall at noon moves almost horizontally westward, with a velocity of more than 18 miles per second, but a stone falling at midnight moves eastward with a velocity slightly greater than at noon.

Note XVII., Page 144.—The author, through an evident inadvertence, stated that " if the earth had had no original rotation, the moon dragging the tidal wave round it 1000 miles an hour (p. 121) would by this time have given it some rotation westward." The tidal wave he refers to is due to the earth's rotation; if the earth were at rest, the equatorial tidal wave would follow the moon in its monthly revolution eastward, at the rate of about 34 miles an hour.

NOTE XVIII., PAGE 162.—Some distinguished observers have seen the crescent shape of Venus with the naked eye. But the crescent, as a symbol of light that regularly waxes and wanes, may have been connected with the representations of Venus, without any actual observations of its literal, as well as figurative, appropriateness.

NOTE XIX., PAGE 190.—If the ring is composed of meteorites, they do not, perhaps, wholly eclipse the sun.

NOTE XX., PAGE 191.—Mr. Denison credited the determination of the structure of Saturn's rings to Mr. Maxwell, a claim which, I believe, Mr. Maxwell himself has never made. Prof. Peirce's results were published in the proceedings of the American Scientific Association, and Mr. Maxwell's essay is, on its very face, a prize essay written to confirm those results.

NOTE XXI., PAGE 218.—Prof. Newton proved that the meteors could have but one out of five well marked orbits, and stated that a computation of certain perturbations ought to decide which one of the five was the true one. Prof. Adams made the computations, and obtained one of the neatest demonstrations possible that only the 33¼ years period could answer the conditions of the problem. [See Monthly Notices of the Royal Astronomical Society (vol. 27, p. 247) for April, 1867.]

NOTE XXII., PAGE 256.—A body colliding with a larger one might, possibly, describe an angular orbit. But the angle also is one of the conic sections.

NOTE XXIII., PAGE 260.—A comet, after leaving the solar system in a parabolic or hyperbolic orbit, might pass around one or more stars in a similar orbit, and finally return to our system. Such an occurrence would, however, be extremely impossible.

NOTE XXIV., PAGE 263.—Professor Peirce has shown that this apparent rapid increase of tail may have been owing to an optical illusion.

Note XXV., Page 266.—The most satisfactory explanation of cometary phenomena has been given by Professor Peirce, in a communication presented to the French Academy, and published in the Comptes Rendus.

Note XXVI., Page 282.—Struve, and other astronomers, have supposed that a large portion of the light which is emitted from the stars is lost by the interference of the various luminous undulations with each other.

Note XXVII., Page 299.—The author has overlooked the physiological considerations which are involved in the question. It may well be doubted whether a man with two eyes can see 2 times as far as a man with one.

Note XXVIII., Page 339.—The first suggestion of the chronograph, for employing the magnetic telegraph to record astronomical observations, was probably made by Professor A. Dallas Bache. Professor Sears C. Walker planned an instrument which was constructed by Dr. John Locke of Cincinnati, and attached to a clock in the observatory of the Philadelphia High School. Professor Wm. C. Bond of Harvard University, afterward invented the Spring Governor, which has effected a marked improvement in the accuracy of minute observations.

INDEX.